ABOUT THIS SERIES...

The ultimate responsibility for determining the fate of this nation rests with its citizens. Given that responsibility, citizens must inform themselves so they can make wise decisions. For the citizenry to do so requires, however, that sources of the facts and knowledge essential to the decision-making process must share in the responsibility by making information readily accessible to the public.

Because of the complexity and seriousness of the problems facing our nation today, special public communications efforts are needed more than ever before. This series represents one such effort by a basic industry, the nation's electric utility companies, through the Edison Electric Institute.

"Decisionmakers Bookshelf" seeks to provide to the public important discussions and reasoned viewpoints on national policy problems related to energy.

ABOUT EEI. . .

Edison Electric Institute is the association of America' investor-owned electric utility companies. Organized in 1933 and incorporated in 1970, EEI provides a principal forum where electric utility people exchange information on developments in their business, and maintain liaison between the industry and the Federal Government. Its officers act as spokesmen for investor-owned electric utility companies on subjects of national interest.

Since 1933, EEI has been a strong, continuous stimulant to the art of making electricity. A basic objective is the "advancement in the public service of the art of producing, transmitting and distributing electricity and the promotion of scientific research in such field." EEI ascertains factual information, data and statistics relating to the electric industry, and makes them available to member companies, the public, and government representatives.

NUCLEAR POWER IN AMERICAN THOUGHT

© 1980 Edison Electric Institute
1111 19th Street, N.W.
Washington, D. C. 20036

Printed in the United States of America
at Sheridan Printing Company, Inc., Alpha, NJ

ISBN 0-931032-08-3

iv

CONTENTS

INTRODUCTION

The outcome of "the nuclear power debate" will turn not on technical considerations alone, but also on arguments involving ethics, psychology, history, philosophy, politics and similar humanistic concerns which affect each of us in ways yet to be fully examined or comprehended.

Nuclear Power in American Thought, a collection of four essays, attempts to shed light on such non-technical concerns and how they affect and may continue to affect the way Americans regard nuclear power.

Edison Electric Institute is pleased to present this volume in the hope it may contribute to the better understanding of nuclear power and in so doing help us better to understand ourselves and the world in which we live.

I

NUCLEAR POWER AND NATURE: INTELLECTUALS AND ENGINEERS

Andrew Hacker

This essay has a single purpose: to trace the basic sources of opposition to the creation of nuclear power. Hostility to this form of energy is strong and deeply-seated. Moreover, it gives every sign of persisting into the foreseeable future. Even if there are no catastrophic accidents in or around nuclear plants. Even if suitable solutions are accepted for disposing of nuclear wastes. Even if the costs of other fuels keep on escalating. Even if inflation and unemployment remain at abnormal levels. In short, antagonism towards nuclear power bids fair to continue even if every effort is made to answer these and other objections.

Concern over nuclear power is more than a political issue. It has also become a vehicle by which a substantial number of Americans express a personal philosophy. In common with all such postures, it can only be understood in its historical setting. The term "historical" has a dual meaning. First, it refers to the experience of the past. Accordingly, Part I of this essay will describe a debate which took place close to two centuries ago. The issue was whether America should embark on an undertaking which would, with the passage of time, bring the advent of nuclear power.

But "history" also means the present. For every era is a period with its own quality and character. Therefore, Part II will describe developments in our own time which have crystallized public sentiment as it applies to nuclear power. Part III will synthesize the

1

strands of this analysis, relating the positions people hold on nuclear power to efforts on their part to define their own identities.

This essay will follow the format of a liberal arts curriculum. The reader will first take two introductory courses: one "history" in the conventional sense; the other "sociology", in that it deals with the generation to which we belong. Only with these requisites completed, will the reader be ready to apply his understanding to the problem of nuclear power.

I

The debate over nuclear power is not just "another" controversy. To be sure, we seem to live in an age that turns everything into an issue. At times the list seems endless, ranging from why our kids can't read to how to cope with crime. It would be easy enough to give nuclear power a place on this agenda. After all, the subject has two sides. Experts can be found, ready to confuse the public with conflicting sets of data. Accidents get front-page treatment in every form of media. There has even been a movie, so no one is really sure at what point fiction ends and where the facts begin.

Even so, this essay will argue that nuclear power is not like the other issues which compete for our attention. The concerns it arouses have a longer and more venerable history than most Americans realize. Outlooks on the question are deeply embedded in the American mind. Indeed, the debate itself began at the birth of this Republic. That confrontation should be recalled, not for its antiquarian interest, but because underlying assumptions were made explicit with a style and clarity we seldom hear today.

The question, simply put, was what manner of nation the new America was to be. While the Founding Fathers were obviously occupied with designing a Constitution, they retained a keen awareness of its philosophical setting. High school students have been taught that the conflict at the time was between those supporting a "strong central authority" and those who wished a government of "limited powers". While that conflict was altogether real, it was accompanied by another division with deeper implications.

That issue, simply put, was whether America should become an industrial society. But the basic issue went further. To be decided was the kind of relationship the men and women of this nation should maintain with their natural surroundings. The debate was especially appropriate for America, then an unopened continent holding a bounty of resources unrivalled in that age. In the Republic's early years, land was assumed to be the basis of both life and labor. Nine

2

out of ten families lived and worked on farms. They had larger and more fruitful holdings than they had ever known in Europe. Ownership engendered pride, as did a living standard of promising proportions. America saw herself as an agricultural nation, and the preponderant sentiment was to keep it that way.

Yet one man at that time had a radically different vision. His name was Alexander Hamilton, then Secretary of the Treasury in George Washington's first administration. Shortly before Christmas in 1791, Hamilton submitted a paper to his President, prosaically entitled, *A Report on the Subject of Manufactures*. Far from being an ordinary memorandum, it was in fact a manifesto with a revolutionary message. Hamilton realized that he was addressing an unreceptive audience. Washington himself was a Virginia planter, coming from an economy which used a form of labor going back to biblical times. So Hamilton began by explaining his ideas in very simple terms. He described, for Washington's benefit, just what machinery was and how it did its work:

> The employment of machinery forms an item of great importance in the general mass of national industry. It is an artificial force brought in aid of the natural force of man; and, to all purposes of labor, is an increase of hands, an accession of strength, unencumbered also by the expense of maintaining the laborer.

To call machinery an "artificial force" makes a kind of sense. At the same time it may be allowed that every machine, from the most primitive to the most sophisticated, is composed of elements which have their origin in nature. Human minds and hands extract these basic components and, by techniques of man's devising, build things we call machines. A machine is "artificial" only insofar as man—himself a natural being—has intervened to combine resources he manages to find in nature. This is not a puzzle over what words mean but an everpresent dilemma over how we act towards our surroundings.

What Hamilton was asking was that this country dedicate its talents to creating methods for combining the ingredients of nature in new and expansive ways. For Washington's benefit, he called machinery an "auxiliary". He then proceeded to phrase his next argument in rhetorical form:

> May it not, therefore, be fairly inferred that those occupations which give general scope to the use of this auxiliary contribute most to the general stock of industrious effort and, in consequence, to the general product of industry?

Here Hamilton was calling not only for new modes of production,

3

but new employments as well. He was explicit in stating his desire to encourage "occupations which give greatest scope to the use of this auxiliary." He had in mind inventors and scientists; but more particularly engineers. If these professions had only a fledgling status in that era, Hamilton wished for them a status yet to be envisioned anywhere in the world. What would make America unique was the honor it would accord to its engineers.

Hamilton held a buoyant view of man's talents and capacities. Moreover he felt that only in an industrial setting would this potential find expression. He wanted Americans to advance as an active, creative people. While there are many avenues for deploying human talent, the one he chose to encourage was industrial productivity. ("To exert imagination in devising methods to facilitate and abridge labor.") Only with that commitment well established would there follow the comforts and pleasures which make for a civilized standard of life and the cultured use of leisure.

So to call Hamilton a "materialist" could be a grave misnomer. While he wished to build a wealthy and powerful nation, it was not as an end in itself. His first concerns were human; "to cherish and stimulate the activity of the human mind." He expanded on this theme:

> Every new scene which is opened to the busy nature of man to rouse and exert itself is the addition of a new energy to the general stock of effort.

> It is a just observation that the minds of the strongest and most active powers fall below mediocrity, and labor without effect, if confined to uncongenial pursuits. And it is thence to be inferred that the results of human exertion may be immensely increased by diversifying its objects.

> When all the different kinds of industry obtain in a community, each individual can find his proper element, and can call into activity the whole vigor of his nature.

An industrialized society would provide a wide range of occupations from which each citizen could choose a calling congenial to his talents. Hamilton wanted America to be a busy, varied place. More than that, he emphasized that each person be spurred to "find his proper element". This is a goal eminently respectful of variant and individuality, and an indispensable accompaniment of the Hamiltonian program.

However there is an underlying assumption here, and it should be made explicit. In the Hamiltonian view, advancing the quality of life, and fulfilling man's potential, requires as its continuing endeavor the transmutation of nature. In this sense, admittedly,

Hamilton was a "materialist". He did not see the good life as possible if men accepted their natural surroundings with only bare adjustments. For Hamilton, the raw ingredients of nature took on the form of a challenge. The processes by which one permutes, combines, and refines these elements are the testing ground of man's creative capacities. Engineering is thus artistry. Indeed it is the art upon which other arts can build. Poetry and philosophy, to reach an audience, must have printing presses. Theaters and museums require subtle shades of lighting. Music can only be reproduced with skillful amplification.

The Hamiltonian plan took further steps as well. Its author realized that for industrialization to be successful, it would require a population attuned to its tempo. Hence his proposal that those engaged in agriculture become part of the new productive modes. His vision was of factories dotted across the country, so that when a farmer had free time he could ride into town and sign on for several hours:

> In places where manufacturing institutions prevail, besides the persons regularly engaged in them, they afford occasional and extra employment to industrious individuals and families, who are willing to devote the leisure resulting from the intermission of their ordinary pursuits to collateral labors, as a resource for multiplying their acquisitions or their enjoyments.

If asked if he desired to depopulate America's rural reaches, Hamilton would doubtless admit that this was his eventual aim. Agriculture as then constituted was grossly inefficient, using too much human labor in relation to its yields. (Among the first industrial priorities would be farm machinery. With the gain in productivity it would bring, fewer Americans would be needed on the land.) Hamilton speaks of his moonlighting farmers as "industrious individuals", encapsulating the temperament he intended to encourage. And he wanted this not only for the farmer, but all members of his household:

> The husbandman himself experiences a new source of profit and support from the increasing industry of his wife and daughters, invited and stimulated by the demands of the neighboring factories.

And if asked if he thought putting women to work might weaken the fabric of the family, his reply would undoubtedly be that women, too, should not be "confined to uncongenial pursuits"—which is how not a few tend to see themselves if they must shoulder all the domestic duties. Women no less than men should be allowed to find

5

"their proper element" and "call into activity the whole vigor of their nature". While Hamilton never elaborated on his ideas for a suitable family life, it would seem he was prepared to welcome arrangements not altogether dissimilar to those we see today. Such losses as would be incurred by mothers' absence from the home would be more than compensated by the talents that would surface from industrial occupations. In short, Hamilton granted that the advent of industry would change the character of the family. He was prepared to take such a risk, in this as in other areas.

He was also ready to transform the character of the culture. Hamilton welcomed immigrants, especially those already initiated in the arts and attitudes associated with industry:

> It is in the interest of the United States to open every possible avenue to emigration from abroad . . . . Factory workers who, listening to the powerful invitations of a better price for their labor, would probably flock from Europe to the United States to pursue their own trades or professions.

> Whoever inspects with a careful eye the composition of our towns, will be made sensible . . . . to a large proportion of ingenious and valuable workmen in different arts and trades, who, by expatriating from Europe, have improved their own condition, and added to the industry and wealth of the United States.

Hamilton asked us to open our doors to anyone committed to the industrial calling. He was not worried lest an America then of preponderantly English origins be joined by arrivals from sundry other nations. If pressed, he would surely say that we should embark upon creating a new industrial culture, one that transcended nationalities and which would draw on capacities all humans hold in common. The factory and the forge would become our melting pot. Blueprints and specifications bespeak a universal language.

Allied to the industrial mentality is an opening of new horizons for opportunities in consumption. Earlier Hamilton had spoken of individuals' putting in additional hours to gain extra income for "multiplying their acquisitions or enjoyments." He returned to this theme:

> The multiplication of factories not only furnishes a market for those articles which have been accustomed to be produced in abundance in a country, but it likewise creates a demand for such as were either unknown or produced in inconsiderable quantities.

The processes of industry in fact create desires which had not hitherto existed. By devising new products—and, in turn, new experiences—such an economy raises the sights of its citizens as consumers. Not only in our work would we find our "proper element", but through our purchases we also express our personalities. A market is not simply economic; from an array of goods and services, each person can confect mixtures suited to his or her identity. And that everwidening range can evoke interests and activities never previously contemplated. Thus to be a "consumer" can be broadening, indeed creative. Like "materialism", a fondness for consumption is easily criticized. Even so, Hamilton would ask us to admit that not only do we enjoy amenities; they also open opportunities for expanding and improving the scope of human life.

There is still another, consequential, premise to the Hamiltonian idea. This is that nature—and man himself—has the strength to withstand its continued transmutation. Answering the Hamiltonian call, America's engineers have created new hosts of compounds and made them part of our environment. These range from simple mechanical contrivances (such as the mouse trap) to chemical catalysts (like preservatives) on to biological breakthroughs (oral contraceptives) and physical transformations (nuclear power).

At this point, fundamental questions arise. How much transmutation by the hands of man can nature itself absorb? May the compounding of these interventions so alter man's surroundings as to threaten his survival? Hamilton, it may be said, was not worried by this issue. He wrote at the threshold of modern technology. His plea was that of an innovator: let industry be given a chance. Indeed, the occupations he had most in mind were those we now associate most with civil and mechanical engineering. These focus more on modular construction than acts of transmutation. The compounding of chemicals and manipulation of physical matter belonged to a stage yet to come. Even so, the prospect that such a phase would ultimately appear was very much on the mind of one of Hamilton's contemporaries.

Thomas Jefferson, in this as in other areas, took a position that was the antithesis of Hamilton's. He would do everything in his power to forestall an industrialized America. For him the health of the nation required that it remain a community of farmers. His assumptions were clear and to the point, and can suffice with a somewhat briefer treatment. He stated them straightforwardly in his *Notes on Virginia*:

When I look around me for security, I find it in the widespread of our agricultural citizens: in their unsophisticated minds, their independence, and their power to maintain the principles which severed us from England.

7

Cultivators of the earth are the most valuable citizens. They are the most vigorous, the most independent, the most virtuous. And they are tied to their country and wedded to its liberty and interests by the most lasting bonds.

Those who labor in the earth are the chosen people of God, if ever He had a chosen people, whose breasts He has made His peculiar deposit for substantial and genuine virtue.

Generally speaking, the proportion which the aggregate of the other classes of citizens bears in any state to that of its husbandmen is the proportionof its unsound to its healthy parts, and it is a good enough barometer whereby to measure its degree of corruption.

The emphasis here is explicitly moral. (Hamilton's also had a moral base; but with a different set of assumptions, couched in a different style.) Jefferson even uses an arithmetic analogy to illustrate his point. If one state counts 90 percent of its population as farmers, while for another the figure is 30 percent, then the former is three times more likely to chose virtue over corruption. Occupations other than agriculture pervert human character, instilling attitudes and values inimical to our nature. For this reason, Jefferson went so far as to propose that no factories be allowed, lest they lure farmers off the land:

While we have land to labor, then, let us never wish our citizens occupied at a workbench . . . . For the general operation of manufacture, let our workshops remain in Europe.

It is better to carry provisions and materials to workmen there than to bring them to the provisions and materials, and with them their manners and principles. The loss by the transportation of commodities across the Atlantic will be made up in the happiness and permanence of government.

Once again, Jefferson expresses his position with an arithmetic equation. Of course it will be expensive to send our raw materials 3,000 miles across the Atlantic for processing in Europe. And it will be equally costly to return the finished goods by the same arduous route. Yet this is just what Jefferson proposed. He knew of Hamilton's plan for establishing factories in every country town. He knew that building a textile mill a few miles down the road from where the cotton was grown would reduce the price of finished products and encourage their consumption.

But Jefferson also realized that this eventuality would induce farmers to abandon the land for industrial employments. And it would also induce the arrival of those "ingenious and valuable workmen" in whom Hamilton took such pride. Jefferson was less impressed with their ingenuity than their "manners and principles," which he viewed as morally deleterious. If America stayed agricultural, such individuals would remain in Europe, a contaminated continent by every moral measure.[1]

And as the costs of transportation would raise the price of finished products to almost prohibitive heights, Americans would find themselves barred from purchasing all save the most minimal of manufactured goods. This too was what Jefferson wanted. An emphasis on consumption could only weaken character, diverting individuals from resources they had within themselves or which were freely available in their natural surroundings.

This view was well expressed in Jefferson's plan for education. While he wanted the new nation to have colleges and universities, he would ask them to place the study of agriculture at the core of their curriculum:

> In every college and university, a professorship of agriculture might be honored as the first. Young men closing their academic education with this, as the crown of all other sciences, instead of crowding the other classes, would return to the farms of their fathers, their own, or those of others, and replenish and invigorate a calling now languishing under contempt and oppression.

Obviously Jefferson believed in learning. Indeed everything about his outlook stressed the need for informed, inquiring citizens. But truth and understanding must emerge in a settled context, if facts and explanations are to have inherent meaning. To say that the study of agriculture should "be honored as the first" asks that man's natural surroundings be taken as the setting in which the intellect can truly thrive. History, art, and literature can only gain significance by reflecting a coherent set of values amid a natural way of life.

Jefferson had alluded earlier to the "unsophisticated minds" of his agricultural citizens. He meant this as a compliment. Unsophisticated is not stupid. (Nor is sophisticated necessarily intelligent.) Simplicity in the intellect is the surest route to truth:

> State a moral case to a plowman and a philosopher. The former will decide it as well, and often better, because he has not been led astray by artificial rules.

---

[1] Thus, in a related connection, Jefferson wrote: "Why send an American youth to Europe for education? He acquires a fondness for European luxury and dissipation, and a contempt for the simplicity of his own country . . . .

"He is led by the strongest of all the human passions, into a spirit for female intrigue, destructive of his own and other's happiness, or a passion for whores, destructive of his health; and, in both cases, learns to consider fidelity to the marriage bed as an ungentlemanly practice, and inconsistent with happiness.

"An American, coming to Europe for education, loses in his knowledge, in his morals, in his health, in his habits, and in his happiness."

Sophistication can mean pretentious methodologies, and vocabularies overloaded with unneeded technical terms. Indeed, there can arise a "technology of learning" which blinds us to reality by imposing barriers against the exercise of common sense. All of us, Jefferson would argue, are natively intelligent. Our task is to create a community in which those capacities will thrive. To achieve that goal, we must acknowledge that we are natural creatures who can develop fully only if we identify with surroundings to which we bear an integral relation.

Thus suffusing all of the Jeffersonian vision is the idea that we should live and labor close to the land. We emerged from the womb of nature and still belong to her bosom. Accordingly, we should come to know her ways and adapt ourselves to her laws. In particular, we should show the greatest hesitation before altering her processes by interventions of our own. The machines Hamilton so admired are in fact presumptuous, evoking overtones of arrogance. They approach nature as if she were a stock of resources, created solely for the benefit of a single species.

So, for Jefferson, machinery takes a toll on nature which can never be redeemed. Even civil and mechanical construction alters the terrain, with consequences going well beyond our immediate comprehension. Chemical and physical processes—not to mention biological manipulation—have an even direr impact, changing the very texture of the planet, and in ways we will never understand.

Not only that, an acceptance of technology transmutes man himself. He begins to view the possibilities of life in light of industrial output. He defines comforts and pleasures in terms of manufactured goods, along with experiences made possible by sophisticated processes. Thus do human beings adapt their minds and spirits to the rules and rhythms of machines. All this Jefferson knew. And feared.

How primal he would have us stay he never made explicit. His simple phrase, "let our workshops remain in Europe", was all he felt impelled to say. His intention in his writing was to establish philosophical priorities. Man is a creature capable of exploiting nature as no species has before. Indeed, he can destroy that which brought him forth. And by so doing he will destroy himself.

This was how America started, with two opposed philosophies based on radically differing visions. In terms of the nation's physical development, the Hamiltonian plan prevailed. The United States became committed to technology as no nation had before. Every part of Hamilton's program was adopted, including his hope that Americans would come to attune their lives to the tempo of an industrial society. It was a success hardly rivalled in the annals of our age. Indeed, during the country's formative century, the Jeffersonian position continually lost ground. Even so, it remained alive, almost as an underground philsophy waiting for a time when the nation would once again embrace its message. In many ways, that hour has now arrived. For if Hamilton's dream succeeded, its very triumph brought a resurrection of the Jeffersonian vision.

II

Contemporary American society owes its basic physical structure to the efforts of its engineers. They are the men (rarely women) who made real the Hamiltonian tradition. They have been our builders, who see the physical world as a challenge, as they seek to combine the elements of nature in new and imaginative ways. This is not the place to explore the relative contributions of science and engineering. Yet if anything, engineering deserves the emphasis because it is concerned less with attaining theoretical understanding than with building things that work. (In fact, most advances in engineering stem less from findings of science than earlier phases of their own art.) At the same time, this physical achievement required a reassuring environment. Corporate industry in America has always honored engineering, for it was the profession on which productivity ultimately depended.

But the progress of productivity had consequences even someone as shrewd as Hamilton could never have predicted. As technology becomes ever more efficient in its use of labor and materials, fewer human beings are needed in the processes of production. Increasing quantities of goods continue to emerge from an ever smaller workforce. The result is that the economy finds itself able to create new kinds of occupations, most of which have only an indirect bearing on physical production; and, in a growing number of cases, no relation at all. This "white collar" explosion could only come into being because advances in productivity engendered the wealth to pay the salaries of the persons occupying the newly-created positions. It would be useful to examine how memberships in the major occupational groupings have changed since the midpoint of this century.

For all practical purposes, 1950 can be regarded as the outset of America's postwar era. The economy was settling into peacetime production. And veterans and other young people were being absorbed into the labor force. In comparing the occupational distribution of 1950 with its present counterpart, we must take into account the fact that the workforce has grown considerably over this

11

period. So to achieve an accurate comparison, the figures for 1950 have been raised to magnitudes comparable to 1979.

Several observations are in order. The growth in "professional" occupations is clearly the most striking. More than six million additional positions have been created in this group alone, surpassing even the lower-paid clerical category. Just what these professionals produce is never easy to say. They may claim, for example, that the activities they pursue augment the aggregate amount of health and knowledge and well-being in society as a whole. At this point all we can observe with confidence is that those in such positions have succeeded in persuading themselves (and many others) that they are doing something which merits drawing a monthly salary check.

The lower growth-rate for the managerial group is at first glance somewhat curious. After all, between 1950 and 1979 all sorts of bureaucracies expanded, especially those associated with governments and other public auspices. A net gain of less than two million managers seems quite modest, inasmuch as the age was said to be bearing the fruits of a "managerial revolution". The short answer is that many people who in 1950 saw themselves as executives (or even more simply, businessmen) now prefer to view what they do in professional terms. One consequence is that the economy now has fewer people who identify directly with business aims and values. Indeed, not a few professionals on business payrolls have a less than complete commitment to the goals of their employers.

The decline of the salesman tells a similar story. One factor is that selling on a face-to-face basis has become increasingly expensive. In some cases (as with supermarkets and discount houses) more of the burden is put on the customer. In others, technology takes over (as with vending machines and automatic order renewals). For the greater part of our history, the salesman stood for the capitalist spirit at its most enthusiastic pitch. Indeed, the vigor he brought to the marketplace was in many ways parallel to the dedication of the engineer. What is important is not just that America has fewer men and women in the field of sales, but the effect this reduction has had on the support business is able to muster in the country as a whole. Its central cheering section has become considerably smaller, even as the population has grown more middleclass by every social measure.

The growth of clerical occupations needs little comment. It simply means, for one thing, that there is a lot more paper fluttering around the economy, supplemented by magnetic tapes and electronic impulses. Certainly, the clerical group has been a convenient haven for millions of women entering the labor force. It may be that professionals more than others like to surround themselves with clerical personnel. For it often seems that professionals have been less than eager to introduce technologies that could cut their supportive staffs. Even with photocopying machines in every alcove, there are as many typists as ever: 3.5 million secretaries at last count.

By contrast, industries engaged in physical production have been willing to adhere to the Hamiltonian dictum, and reduce their labor forces wherever possible. The economy has effectively lost more than 13.5 million blue collar employees, due chiefly to the introduction of more efficient machines. In fact, the reduction is even greater than suggested by the statistics. Among the current category of 47.6 million blue collar workers, many are actually in "service" occupations rather than actual production. They repair, package, deliver, and generally serve a facilitative function in the industrial chain.

The United States has thus experienced a substantial growth in those categories of occupations most removed from physical production. This has had significant effects on attitudes towards technology itself. For the kinds of work which people do influences how they choose to see themselves and the world around them. Of course Hamilton and Jefferson said just this. It happens to be true.

In addition, another "occupation" has come to occupy a promi-

Table One

OCCUPATIONAL DISTRIBUTIONS: 1950 - 1979

	1950		1979	
Professional	4,909,000	9%	15,050,000	16%
Administrative	5,018,000	9%	10,516,000	11%
Sales	3,927,000	7%	6,163,000	6%
Clerical	6,894,000	12%	17,613,000	18%
Blue Collar	35,478,000	63%	47,603,000	49%
	56,226,000	100%	96,945,000	100%

	1950 figures expanded to fit 1979 total		Real net changes 1950-1979	
Professional	8,465,000	9%	+ 6,585,000	+ 77%
Administrative	8,653,000	9%	+ 1,863,000	+ 21%
Sales	6,772,000	7%	− 609,000	− 9%
Clerical	11,888,000	12%	+ 5,725,000	+ 48%
Blue Collar	61,167,000	63%	−13,564,000	− 22%
	96,945,000	100%		

nent position in the lives of many Americans. This is the experience of being a student, especially at the college level. Due to the improvement of technology, the economy no longer requires the services of most of our young citizens. We can sequester them in colleges at no productive loss. Not only that, enough resources are available so that either individual families or society in general can afford to underwrite these expensive institutions. (Not the least reason they are so costly is that they are mainly staffed by well-paid professionals.)

If between 1950 and the present, America underwent an educational revolution, it was made possible by progress in productivity. In a sense, students are novitiate professionals, supported by a system to which they make no contribution, and bolstered by the illusion that in time their services will be needed.

From 1950 through 1978 (the most recent year for which we have statistics) and once again adjusting for population growth, the number of college graduates rose by more than two and a half times, from the equivalent of 7.4 million to 19.3 million. The number of those who attended but did not complete college almost doubled, from 8.8 million to almost 17.4 million. (It is interesting to note that while in 1950 the number of college drop-outs exceeded the graduates, by 1978 that ratio was reversed.) One in six Americans over the age of twenty-five has had at least four years of college. Among those aged 25 through 34, the proportion was almost one in four. The remainder of the population is now more apt to have finished high school than not. (Indeed, it is striking that as recently as 1950, fully two-thirds of adult Americans did not have a high school diploma).

I am not concerned to argue here whether Americans know more due to their added schooling. What is clear is that they believe they have sufficient knowledge to formulate opinions on a vast array of topics. In addition, they feel entitled to voice these sentiments, and they expect that their views will have a discernible influence on what happens in society. Education, especially at the college level, expands the estimates people have of themselves. An increasing proportion of Americans have confidence in their capacity to assess and evaluate what they believe they see around them.

All this needs to be said because education in the past tended to be seen as training. Most people who went to college (and most of course did not) studied such subjects as home economics and animal husbandry, or accounting and pharmacy, with engineering and elementary education leading in enrollments. Pure liberal arts majors comprise a distinct minority.

Table Two

EDUCATIONAL ATTAINMENTS: 1950 - 1978

	1950		1978	
4+ Years College	5,285,000	6%	19,332,000	16%
1 – 3 Years College	6,259,000	7%	17,379,000	14%
4 Years High School	17,664,000	20%	44,381,000	36%
Less	58,277,000	67%	41,928,000	34%
	87,485,000	100%	123,019,000	100%

	1950 figures expanded to fit 1978 total		Real net changes 1950-1978	
4+ Years College	7,434,000	6%	+ 11,898,000	+ 160%
1 – 3 Years College	8,804,000	7%	+ 8,575,000	+ 97%
4 Years High School	24,840,000	20%	+ 19,541,000	+ 79%
Less	81,941,000	67%	– 40,013,000	– 49%
	123,019,000	100%		

Yet this began to change during the 1950's. By and large, subjects once looked upon as "practical" lost their lead, to be replaced by those seen as "theoretical", or even "humanistic". Schools of education, along with agricultural and mechanical institutes were upgraded to university status with more catholic ranges of offerings. It became possible to major in Philosophy at Texas A. & M. Engineering in particular saw its graduates coming to represent a diminishing share of those receiving college degrees.[2] In contrast, a field like sociology grew at a faster rate than overall enrollments.

The sum and substance of these occupational and educational statistics is that people who build things comprise a dwindling sector of American society. Our major growth industries consist of those who make their livings by talking, reading, and writing. Between

---

[2] Nor can it be said that pure science made up for the drop in engineers. Between 1963-1964 and 1973-1974, the number of bachelors' degrees awarded almost doubled. Adapting the 1963-1964 figures to the 1973-1974 total (as was done with the earlier tables) degrees in engineering underwent a 30 percent drop. In the physical sciences (mainly physics, chemistry, and geology) the relative decrease was 36 percent. For mathematics, the downward figure was 39 percent. Only in the biological sciences did the number of degrees grow in relative terms, and there by only 12 percent. Moreover, many if not most of these can be assigned to the pre-medical impetus of the time.

1950 and 1979, the number of journalists in the nation more than doubled, while librarians more than trebled. By 1970, the country had more than four times as many college professors as it had in 1950. The social work profession enjoyed a fivefold expansion; and 1979 found six times as many psychologists compared with 1950.

For purposes of this part of the analysis, I have decided to divide those most involved in the nuclear debate into two admittedly arbitrary groupings. To oversimplify even further, I have chosen to call them "engineers" and "intellectuals". Any such division runs all the risks attendant on any stereotyping. By "engineers" I mean those who tend to hold a sympathetic attitude towards technology—an outlook shared by many people who have not had technical training. By "intellectuals" I refer not only to individuals absorbed in scholarly ideas, but a rather larger group who wish to detach themselves from physical production.[3] In any case, the reality under discussion consists of sentiments which, in varying degrees, can be found in all Americans. They may be best understood when expressed at their most intense, which will be the format here. For how these sentiments add up will determine the ultimate outcome of the nuclear debate.

The engineers can be counted as "new" Americans to the extent that they, too, tend to see themselves as professionals and are apt to be well educated. However, they also represent the older tradition which sees its task as transforming the physical world. In this, they represent today's adherents of the Hamiltonian vision. By and large, contemporary engineers have integrated themselves into the corporate structure. Most work for industry, seeking to improve productivity as a means of enhancing employers' earnings.

However, a problem for the engineers as a group is that they have never been articulate. In the past this did not matter much as there was an American consensus on the value of technology. Who could possibly object to innovations that would prolong life and health, increase leisure and enjoyments, create new employment, and solve so many of the problems that arise in any society's midst?

---

[3] I am not altogether happy about the use of "intellectual". For one thing, it is often employed as a term of derision: referring to radical professors, avid for a greater governmental presence in all aspects of our life. My own focus is on a wider constituency: perhaps best characterized as those still bearing the effects of their exposure to a liberal arts education. Other dichotomies I considered were "pragmatists" and "idealists"; "realists" and "romantics"; and, of course, Hamiltonians and Jeffersonians.

III

Yet now their works have come under criticism. And by the very individuals whose educations and occupations were made possible by the engineers' achievements. This is of course an irony; but it is also part of that process we call history. Every economic system, every mode of technology, unwittingly engenders forces subversive of its premises. That America's agricultural economy gave rise to an Alexander Hamilton exemplified the process. Yet it was not Hamilton's writing alone that brought the change, but that farming itself had reached the point where those engaging in it were ready to be lured to industrial occupations. By the very same token, Hamiltonian technology has reached a point at which it is creating people who react as intellectuals rather than engineers. One need not declare as dogma that every system sews the seeds of its own destruction. It will suffice to say that America has many people in its midst—more than ever before in this century—who now oppose the basic principles of a technological society.

Does this mean that the values of Thomas Jefferson are enjoying a revival? Clearly, no one is arguing for a full "return" to the past, anymore than people can or will go "back to nature" once they have left the forest. Still, it is accurate enough to say that the intellectual stratum is basically Jeffersonian. Needless to say, they have no desire to devote themselves to farming, at least as a means of fulltime livelihood. Rather, intellectuals see themselves as representing the coming phase of history: a post-industrial epoch. In this they are, by their own estimates, superior to the engineers. As evidence of this, they will allude to their greater sensitivity in appreciating how human life has been affected by the forces of technology. The intellectual views the engineer as committed to exploiting nature regardless of the consequences. To this perception, the engineering persuasion lacks a philosophical perspective. (One never hears its case in Hamiltonian cadences.) The engineer is accused of never pausing to wonder whether nature can stand technological intervention on so vast a scale. The intellectual feels he has the distance and the leisure, aided by his verbal facility, to contemplate these questions. That, after all, is what liberal arts is all about. He has decided that the Hamiltonian era has run its course.

It has become customary to refer to "environmentalists" as if they were a small group who get their way by strategic legal challenges. Certainly associations such as the Sierra Club have minute memberships. Still, overall environmentalist sentiment is a broad and pervasive presence in the United States today, even if most who

17

share those views, neither have formal affiliations nor engage in public stands. There has been a "return to nature" in the sense that more Americans see themselves as having a philosophical interest in preserving not only their natural surroundings but a conception of themselves reflected in such a view. Thus technology is seen as something alien: a predator whose very processes undermine the basic balance of life. Thus nature is viewed as superior to those seeking to transmute it. The intellectual sees himself as standing with nature, against the insensitive engineer. In this he casts himself as Jefferson's rightful heir.

Outcroppings of this attitude are not difficult to detect. For example, the superiority of cross-country skiing, contrasted with snowmobiles. Or the symbolism of plain wood floors, as opposed to acrylic carpeting. Or stir-fry cooking in a wok, rather than switching on a microwave oven. Or demanding that one's food contain nothing aritificial. Outlooks such as these are far more likely to be found among social workers and librarians with liberal arts degrees, than chemical engineers from the other end of the campus.

To be sure, engineers do not deny themselves the amenities of life. After all, they seem quite capable of spending the incomes that come their way. Still, if asked, they will argue that their purchases involve less in the way of material goods than uplifting experiences. Rather than a motor boat, it will be a trip through rural France. Instead of a swarded suburban home, there will be a vacation cabin in the woods. Not so much a showy wardrobe as season tickets to the opera. Of course it may be rejoined that it takes some sophisticated technology to airlift an intellectual to Europe, or to orchestrate the lighting for a production of "Aida". Even that cabin in the woods will want an electrical line. And those skis for country-crossing may be molded from fiberglass.

Even so, the idea that the purchasing of experiences stands at a higher—and more natural—level than acquiring tangible goods goes hand in hand with the desire to declare one's self independent of technology. Perhaps the Jeffersonian "simplicity" of modern intellectuals is more a moral symbol than a sign of willingness to reorder one's whole life. If so, the Hamiltonian persuasion still maintains an unacknowledged hold. But not enough to allow the intellectual to support so ultimate a technology as nuclear power.

Nuclear power, then, is seen as an ultimate transmutation of matter. To those who regard themselves as heirs of Jefferson, it is man's manipulation of nature expressive of an arrogance unknown in earlier technologies. Indeed, it is viewed as an unnatural act, defying laws and precepts our species would do well to obey. Needless to say, the use of nuclear power cannot help but have been affected, indeed tainted, by its military applications. Nuclear power was originally a technology of war, one still the most terrifying of weapons in our nation's arsenal. Its very presence symbolizes a line we should never have crossed. Just as we must never "use" nuclear weapons, so from this it follows we should not draw on nuclear power in any form. Just as Prometheus took fire from the gods and was punished for all eternity, so nature will see that we suffer for a parallel transgression.

If the engineer sees the physical world as an arena for adventure, the intellectual is much more fearful of unknown consequences. There is even anxiety lest some kind of "chain reaction" be set in motion, causing earthquakes or tidal waves, or so contaminating the atmosphere that life itself will be in jeopardy. Concern over radiation bespeaks a more general dread of unleashing lethal vapors we cannot see or smell or feel. Until it is too late.

Attitudes about nuclear energy relate to still another set of circumstances. It would be well to take a moment to recall the underlying basis of opposition to the war in Vietnam. It was that our intervention in Southeast Asia represented a purblind use of power against a people far more primitive than ourselves. Protests against the military aspect were only part of the indictment. More fundamental was the fact that we had become a corporate state: a technological complex whose reflexive mode of response was systematic destruction. Opposition, therefore, was not simply to the war itself. It was to what America had become. The plea "Come Home, America," was not merely asking that our troops withdraw. Rather, it voiced the hope that the nation as a whole would return to an earlier, gentler way of life. It was an entreaty to America to rediscover its Jeffersonian past.

In that sense, the opposition to that "war" is far from over. Reactions to nuclear power express feelings of hostility to the nation's corporate structure. And nuclear power is "corporate" by every modern measure. Electrical utilities themselves are seen as impersonal structures, staffed with administrators and engineers who are bound by institutional imperatives which cannot transcend technical interpretations of the general good.

Quite plainly, opposition to nuclear power is "political" in character. But it would be a grave mistake to view that confrontation in conventional terms. Opponents are not necessarily on the "left", nor do most perceive themselves as socialists or radicals in any formal sense. Few call for government ownership—if only for the reason that there is little faith that government would do things any differently. (The Tennessee Valley Authority's interest in nuclear power provides evidence on this point; so does the power authority owned by New York State.) If opposition opinion seems to be "anti-business" in tenor, it is only because most utilities happen to be privately owned. Were they to come under government ownership, the protests would still continue.

The feelings outlined here are essentially romantic. America's corporate structure is the institutional analogue of its technological twin. Impersonal organizations are seen as social machines, run by the same rules as the cooling towers they build. (A socialist state can be just as corporate as its capitalist counterpart. Hardly any opponents of nuclear power admire the Soviet Union.) If anything, the Jeffersonian intellectual would like to see a dismantling of the entire corporate state—a complex seen as linking both government and business, research and education, even much of medicine and certainly the military. This is the heart of the message in the rock music concerts opposing nuclear power. One cannot understand this sentiment without listening to the lyrics of Carly Simon and James Taylor, John Hall and Jackson Browne, and Crosby, Stills, and Nash.

Of course such an outlook may be called "naive". It certainly seems "unrealistic" by engineering standards. If pressed, intellectuals will speak of solutions which have a "natural" aura. Hence allusions to solar energy ("capturing the sun") and the seashore ("harnessing the tides") or even burning wood (so long as we replace it). To point to problems of cost or feasibility does not come to grips with the romantic impulse. It was this order of feeling which led Thomas Jefferson to say that all the expense incurred in sending goods twice across the Atlantic would "be made up in the happiness of government". This is what happens when adherents decide to discuss issues in purely moral terms.

In this connection, too, it may be added that what we called the "Sixties" are not entirely over. Of course, overt radicalism has generally subsided. On the surface, most young people seem ready to enter the system as it is currently constituted. The Sixties generation came out of the prosperity of that period, and could afford the luxury of a purely moral posture. (And, as was indicated earlier, the freedoms enjoyed by these youthful citizens stemmed from the advances wrought by engineers within the corporate structure.)

However, even while preparing themselves for careers—which they prefer to see as "professional"—young people carry a measurable degree of detachment from the system they will enter. We can note that while the generations of the Seventies and the Eighties dutifully do their lessons, they lack the loyalty and dedication which marked the corporate recruits of the Eisenhower

20

Fifties. Today's young people combine an awareness of the requirements of a Hamiltonian structure with a substratum of Jeffersonian romanticism. In the morning they may march off with their attache cases and the *Wall Street Journal*. At night, they may light up and turn on John Hall's "Plutonium is Forever."

So the opposition will continue, quite apart from whether the nuclear industry solves the problem of wastes and avoids a catastrophic accident; even if fuel prices continue rising and unemployment persists. The advent of nuclear energy, as was said at the outset, is more than just another issue. It has brought to the surface basic judgments about the lives people want to lead. To be sure, not many individuals are inclined to state their feelings as explicitly as this essay. Nor are more than a handful aware that their ideas have antecedents going back two centuries. But, as Jefferson remarked, one need not be a professor to participate in a debate of philosophical proportions. The purpose of this essay has been to emphasize that this is what we are observing.

II

NUCLEAR PHOBIA: PHOBIC THINKING ABOUT NUCLER POWER

Robert L. DuPont, M.D.

At the invitation of the Media Institute, a non-profit organization devoted to improving the quality of media coverage of business and economic affairs, I recently had the unique experience of watching over 13 hours of TV news videotapes on two consecutive days. These tapes contained all the news stories regarding nuclear power broadcast by ABC, CBS and NBC on their evening news programs between August 5, 1968, and April 20, 1979—one month after the Three Mile Island accident.

An earlier study by the Media Institute reported many quantitative aspects of this fascinating tape file—including the fact that 42% of all the network news coverage of nuclear power during this 10-1/2 year period occurred during the last 30 days—the TMI episode.

Having reviewed the videotapes, my impression is that fear was the motif of the entire series of nuclear news broadcasts. The driving force in the nuclear energy issue is not economics, not technology, not energy production, it is fear. There is no question in looking at these TV tapes that the public debate is primarily going to hinge on

ACKNOWLEDGEMENT

Dr. DuPont's essay, "Nuclear Phobia: Phobic Thinking About Nuclear Power," was published previously by The Media Institute, Washington, D.C. EEI is grateful to The Media Institute and to Dr. DuPont for permission to present the essay in revised form.

the fear and that, right now, the tide appears on the tapes to be running with those who are encouraging the fear.

In fact, those who are reassuring about nuclear energy are themselves being consumed by the dynamics of our public fear of nuclear energy and undermined as being self-interested, as being part of the nuclear energy industry. Ralph Nader said essentially that nuclear power is the nation's technological Vietnam and that the government and the nuclear experts have a vested interest in promoting nuclear energy. This implies that almost anybody who knows anything about nuclear energy has an investment in it and therefore cannot be believed, unless, of course, he is among that group who are saying nuclear power is as bad as you think, or worse. Only then, on the tapes, did an expert appear credible.

The paradox that a person like myself, or anyone connected with the media, has to deal with is that real social progress can result from fear. Progress never comes without paying a price. The new regulations that are going to come from the 1979 airline crash in Chicago are going to be paid for by all of us in higher air fares. They may or may not reduce the risk of flying. The same thing now appears to be happening with nuclear energy. The cost of diminishing the hazards may make nuclear power uneconomical. But economics are not the driving power behind the nuclear power debate. The debate is hinged on fear of a particular kind. I call it nuclear phobia, or more precisely, phobic thinking about nuclear power.

A phobia by definition is fear based on exaggerated, unrealistic danger. For example, if you are afraid when a lion bounds out of a cage at the zoo and comes toward you, that is normal, healthy fear: the danger you face is external and substantial. If, on the other hand, you are afraid to go into the zoo at any time because the sight of a snake—even in a cage—strikes terror in your heart, that is a phobia. Phobic people are generally mentally healthy people who have been sensitized to particular experiences or situations which trigger terror or panic reactions.

Phobic reactions are seldom an exaggeration from zero risk. Often there is some risk in phobic situations. For example, people who have an elevator phobia say, "The elevator could get stuck." It is true—an elevator could get stuck. And a person who has a fear of flying knows an airplane could crash. Many people suffer from hypochondria, a related condition. A hypochondriac asks, "Could I have cancer?" The physician will have to say, "Yes, you could. My best judgment is that you do not. I have looked you over and have not found cancer, but it could be there." Phobic fear is out of proportion to the real, external danger. Airplanes do crash, and elevators do get stuck, and lions do escape from their cages at the zoo. But in all these situations we normally act as if the risk of these problems does not exist, since the risk is so low. The phobic person, however, despite reassurances to the contrary, experiences these situations as if the risk of danger were high. His individual phobic response is involuntary. In most cases, outside the phobic situation he thinks clearly and realistically. In addition, many, but not all, phobics fear that they will lose control of themselves—that they may leap from a height, make fools of themselves in a supermarket, or pass out. Thus, with most phobias there is an inner dimension—fear of loss of control—as well as an external dimension—fear of the airplane crash.

To put it differently, a phobia is, at its root, a malignant disease of *what ifs*. The phobic thinking process is a spiraling chain reaction, to use an atomic energy analogy, of *what ifs*. Each *what if* leads to another. "*What if* this happens and then *what if* that happens, and then *what if* the other thing happens?" Moreover, phobic thinking always travels down the worst possible branchings of each of the *what ifs* until the person is overwhelmed with the potentials for disaster. Often these worst-case eventualities remain only partly articulated even to the phobic himself. Feelings of dread mix with half-formed thoughts of disaster to produce a phobic spiraling of terror. When I reviewed the network news nuclear tapes, one of the striking characteristics I saw occur over and over was the reporters continually going down those *what if*, worst-case branches.

Consider for a moment The China Syndrome, a Hollywood film about an accident at a commercial nuclear power reactor that predated and presaged Three Mile Island. I talked about it with a person who saw the film. She was terrified. I asked what really happened in The China Syndrome? The answer was, "Nothing—but it almost did—there was almost a disaster." You ask the same question about the Three Mile Island incident and other nuclear incidents. The actual feared thing itself did not happen, yet the frightened person says, "But it almost did." Characteristically, the fear is not of what did happen but of what almost happened. The "it" in this statement itself is a phobic distortion—a condensation into a single pregnant pronoun of a myriad of potential negative scenarios ranging from trivial leakage of small amounts of radiation to nuclear explosions. In phobic thinking, the event as it actually occurred is rarely feared. The *what ifs* are feared. Phobic fear is a fear of fear; fear of the panicky feeling.

None of this is to say that everyone who opposes nuclear power has a phobia. Nor am I saying that opposition to nuclear power is wrong—or, for that matter, that support of nuclear power is right. In our Phobia Treatment Program in Washington we have not seen a single person seeking treatment because of "nuclear phobia." Nevertheless, after viewing these network TV news tapes, I am convinced that fear is the dominant theme of this particular TV coverage and that much of the fear of nuclear power has elements of phobic thinking.

Finally, phobic people are generally intelligent people. In fact, the ability to anticipate problems before they occur is the hallmark of intelligence. Thus it would be foolish to deny all possible or *"what if"* problems. The difference between reasonable anticipatory anxiety and phobic thinking is clear at the extremes but blurred at the boundary. Fear of getting on an ordinary elevator is irrational and unreasonable. Fear of motorcycle racing is not so obviously phobic, even if one has never had an accident! Thus the second point: the definition of what is reasonable is, in some cases, a matter of judgment. Not everyone will agree with my point of view about every situation. My observation, however, is that fear based on *what if* rather than what is, especially when this fear involves a spiraling, poorly articulated chain-reaction of potential disasters, is likely to be phobic fear.

To accept nuclear energy then is like accepting the airplane as a means of transportation: not 100% foolproof but generally safe. The airplane is a wonderful example because as a passenger you do not have control. To fly you must put yourself in the hands of the Federal Aviation Administration, the airplane maintenance people, the pilots, the Weather Bureau—all these faceless, unknown people. Essentially you have to give to authorities, people who are charged with the responsibility of caring for you, the responsibility of your life. And you say, "Okay," when you step on that plane. You no longer have the capacity to influence your own safety once that door shuts. There are no parachutes on board. That sounds like a scary thing to do. But for most of us who fly, it is with a sense that there are few things we do in life where there are so many checks for safety as there are on a commercial airplane flight. That does not mean that flying is absolutely safe.

In order to lead a normal life, we all must have a certain amount of trust, to deal with our own *what ifs*. Today, in nuclear energy, this needed trust is eroded and we are left with our own hazy half-formed, *what ifs* about nuclear energy: "Can you be absolutely certain there will not be a meltdown?" "Can you be absolutely certain there will not be an accident?" "Can you be absolutely certain that whatever. . .will not happen?"

You cannot be certain, no one can. The media pushes microphones in the faces of the experts and says, "But can you be certain?" The honest expert will say "No." So it is reported that the head of the Nuclear Regulatory Commission, or whoever the expert is, admits that a nuclear accident could happen. So what else is new in life? Of course an accident could happen. It is a question of balancing risks *versus* the alternative risks and assessing the costs. It is also a matter of probabilities. That perspective was almost never brought out on the tapes; there is almost no mention of risk balancing.

Moreover, the nuclear power industry itself contributes to this feeling of all or nothing. Industry people say, for example, in trying to prove how safe nuclear plants are, that there has never been a death from radiation in a nuclear power plant or in the surrounding communities. That is too much, of course, because there *could* be. The fact that there have been none is a substantial achievement, but there could be an accident and people ought to know that. The best statement I heard on this point was by a scientist who talked about a study in Washington state which reported that of 4,000 deaths from all causes over many years of persons who worked at a nuclear facility, one scientist estimated that there were 30 to 50 "extra" cases of cancer which may have been due to something—probably radiation—at the nuclear plant.

The reporters who covered the story asked questions and used a tone of voice which carried the message: "Well, God, is that not terrible?" The scientist said, "No, this is a risk that comes from that activity. People ought to know the facts, but you should not close down the plant because of that additional risk." While this particular study may be disputed and even ultimately incorrect in exaggerating (or minimizing) the danger to workers at nuclear plants, this is not my point. This kind of straightforward information—including negative potentials—is what people need about nuclear power and alternative power sources. I saw little perspective in these TV tapes about the risks for the workers in the plants or for the public.

Nuclear energy is particularly conducive to producing phobic fear. First of all, nuclear energy is still alien. It is not like elevators, or fire, or lions. Some of its dangers are invisible, and there is also a long, uncertain delay between exposure and resulting health problems. I am reminded of a friend who was outraged about nuclear waste disposal and the hazards it might produce in some future time. While she was expressing her anger, she smoked a cigarette. On the one hand, she worried about a hypothetical danger that sometime in the next 250,000 years there might be seepage from nuclear waste; on the other hand, she had an absolutely certain health danger burning between her lips. The differences were two. She was familiar with and in control of one risk; she was not familiar with and not in control of the other. One large, well-known risk she denied, the other small, hypothetical risk she magnified.

People also fear nuclear explosions. The atomic bomb is fixed in our collective consciousness as an ultimate and terrible explosion that killed tens of thousands of people. Experts say that nuclear reactors cannot explode in the sense of an atomic bomb explosion. But people still think of the image of the bomb. In an era of mistrust of all experts, especially nuclear experts, who can be sure a nuclear plant will not explode like an atomic bomb? Another unique aspect of the nuclear threat is the long term nature of the risk of radioactivity. You know you might survive a blast. But as one of the workers said on the tapes: "What if I get skin cancer in 15 years? Who is going to take care of me then?"

Nuclear fear is thus the playing out of the *what ifs* of our age. It is not like getting off the airplane and feeling, "Whew, I made it off the plane!" The risk of radioactivity is more like getting off the airplane and feeling that 15 years later you *might* die from that plane trip! You cannot tell for sure.

I know a person who has a family history of a disease called Huntington's Chorea. It is a deadly, inherited neurological condition. It shows up only in middle life, in the 30's and 40's. There is no way to predict it. Half the children of the people who have this condition will inherit it. You never know who it will strike, so all the children of an effected person have to live their entire early adult lives, marrying and having children, or rejecting these options, never knowing. They do not know whether or not they have this condition and never know whether or not they have passed it to their own children until they are themselves middle aged. Every time they have an abnormal muscle movement, they wonder if it is the first sign of the disease. It is devastating to have to deal with that mysterious, terrible threat. *"It"* could be there. That is the thing that happens with nuclear radiation, the unknown nature of the risks over many years nourishes the fears in a way that is unique to nuclear fear. The only fear that comes close is the fear of chemical waste contamination and the fear of food additives. These fears share much with the fear of nuclear power.

Of course each new technology has produced fear. Electricity and natural gas are examples as are the train, the automobile and the airplane. None of these technologies was or is absolutely safe. We should continue to make them safer. But none of these now familiar technologies produces widespread fear today.

Nuclear energy is also unique in terms of the public perception that only brainy people can understand it. In addition, nuclear power—from the early experiments of the Curies to the Bomb to the nuclear power plant down the road—has been experienced as the scary and potentially evil genie which ordinary people cannot understand.

For one thing, nuclear power is controlled by powerful, remote, establishment organizations—the government, the energy companies, etc. Many who fear the influences of these powerful people and institutions also fear nuclear power for these reasons. Nuclear power uniquely taps anti-military feelings because of the connections in many people's minds of nuclear power and the atomic bomb. Then there also is a much more primitive concern: it has to do with living and dying, and with one's offspring. Again it is the issue of an uncertain and therefore more terrifying risk.

What we have never done, and what we desperately need to do, is to be able to put the risks of nuclear power into an understandable, familiar context so that we can relate to them. Take the genetic issue. The presumption from seeing these TV news tapes is that in the absence of exposure to radioactivity, one is not going to have genetic problems. On the other hand, if you are exposed, there is an assumption that you will have genetic changes.

In the first place, it is important to realize that genetic changes—or mutations—are part of normal biological life. Man's evolution has depended on genetic mutation, and presumably this is a continuing process. We know that mutations can lead to better health and adaptive responses, just as they can produce genetic disasters, though negative mutations are far more common. Certainly, with increases in radiation levels there is an increased risk of mutation, and most of these will be negative. Notice, this is an increased risk, not a certainty, of genetic mutation. It is the same with cancer. The presumption on the tapes is that you either will not have cancer if you do not get radiation exposure, or you will get it if you do. That is incorrect. The truth is that you have an increased *risk* of cancer with increased radiation exposure.

Finally, the tapes suggest that nuclear power is a primary source of contemporary radiation exposure. This also is incorrect—radiation exposure is part of our environment, both natural and man-made. The additional risk of radiation from nuclear power needs to be seen in perspective. It is unrealistic to even think of achieving a zero radiation exposure. This goal destroys rational discussion. Television as a medium helps transmit phobic thinking in two ways. First, you see people in various roles on camera who are afraid. Then you are told by some experts who are critical of nuclear power, authority figures, that *they* are afraid. Think about that. That was what was so significant, for example, about Howard K. Smith's comments about nuclear power—his bottom line, his last comment after Three Mile Island: "I fear it." He gave the same arguments he had once given in favor of nuclear power, but this time there was a twist at the end. He said, "I fear it." And now he implied, because he had fear, he did not support nuclear power. And that is the key. You see authority figures who say, "I fear it." That is what the child understands when he sees his mother afraid of a thunderstorm. This authority figure, this window on the world, says the right way to behave is to be afraid.

Second, television contributes to undermining our national trust in authority. The people who are saying it is okay to get on the airplane, it is okay to take this risk, are called into question. The industry spokesmen who are interviewed appear as fat cats, special interest groups and scary, brainy types. They are stereotypes, targets to put down, and they are balanced by "consumer interests" or "concerned scientists." In these tapes, the critics seem to have the high ground in terms of being the ones who are saying, "You are right to be afraid!" The presumption is that the critics are less self-interested.

There is a big question about whether they are. The issue has to do with the payoff they get from the media attention itself. They get a tremendous boost to their ego in feeling they are doing something noble, God knows we all like that. They get attention payed to them. It is educational to see the extent to which the anti-nuclear protesters, and the groups that are formed around this issue, are dependent on media attention and are hungry for it. The media response is eerie. The presumption in the media often, but not always, seems to be that the protesters and the other "anti's" are free from the taint of self-interest, in contrast to the experts.

The fear transmitted by the tapes tended to concentrate on three main areas of concern. One was the disposal of the waste, and the standard that seemed to be applied was, "Can you assure us that during the next 250,000 years there will be no leakage or problem from this waste?" Of course, the answer was "No." "Then it is unacceptable," was the implied conclusion.

The second concern was about terrorism and theft. "Can you be certain that there could be no terrorist assault that would either sabotage the nuclear power plant and cause a nuclear disaster or lead to theft of radioactive material?" The answer, again, had to be "No." You cannot say absolutely that every facility is secured at all times from any conceivable attack.

30

The third question was, "can you be sure that there will not be an accident that will lead to contamination of the environment, whether it is a meltdown or just a venting of radioactive water or gas?" Again, the answer was "No"; experts cannot be completely reassuring. Those three dangers come up over and over again. And the inability to say absolutely there is no danger is presented as an admission that nuclear technology is faulty.

The critics of nuclear power seemed to have the high ground on the tapes. There have been and continue to be accidents—unexpected occurrences of some danger. Following each accident an investigation shows mechanical, design, and/or human failures. Even if the accident produces no injuries and even if the rate and seriousness of the accidents are less than in comparable technologies, the occurrence of the accident itself is seen as justification for the critics. Finally, even in the absence of an accident, the critics appear "right" because no one can say an accident could *not* occur.

This kind of thinking prevails because we are missing a sense of perspective. We do not ask that stewardesses and pilots who are exposed daily to cosmic energy radiation be taken out of airplanes. Yet they are exposed to too much radiation over time, according to some standards. Medical radiation from x-rays is another more familar risk. That perspective is almost totally missing from the tapes. In fact, whenever anybody tried to bring up something like that, it seemed his credibility was undermined. Why? Because of the *what ifs*. I know this from my own work with phobics. What if somebody that I have helpe drive a car or return to air travel is involved in a serious or even a fatal accident? People want to avoid being in a responsible position for any bad event. Medical people are especially fearful of being in that position. In addition, we must be realistic; there is also a history of radiation standards that were acceptable to the authorities in previous decades or generations, which are no longer acceptable. The standards have been raised, and this change in standards tends to undermine the public's trust in the judgment of the authorities today.

Most reporters covering the Three Mile Island accident kept a pretty good distance from the controversy. I did not see many who seemed to understand it, or who were involved directly. Their detachment was appropriate. Some, however, looked fearful during the Three Mile Island crisis. Walter Cronkite of CBS looked harried, confused. On the Friday night broadcast on ABC, both Betina Gregory and Max Robinson looked frightened. Part of that apparent fear may have been their inability to figure out exactly what the story

31

was. I would think, in retrospect, that they would have to worry about what that story really was because they would have to know, at some level, that it was a story about *what if*, or what might have been. That is not common in journalism. You just do not see a lot of stories of planes that might have crashed.

All this raises the important question of, "What are we interested in?" And you might ask the question, "What makes the news?" Why is this news? The answer is: Fear is news because fear is interesting. One of the characteristics of the fear reaction is a heightened sensitivity to what is around a person and a heightened sense of what might happen. Psychologically the reason for our interest is that the adrenalin gets flowing, our minds become more alert. It is like taking amphetamines. Fear is an upper. You are ready for action! This upper effect works for the media. The media is drawn into that; it has to be, simply on the basis of the biology. The longest lines at amusement parks are always for the scariest rides.

Most of the people in the media probably have not thought much about what the factors are that push them into those positions. Many are aware that what gets people's attention is fear. If you reassure people and say there is no problem, that is like saying, "There is no news. I do not have anything to say." But if you say, "There is a problem, there is danger," you are going to get people's attention. On an intuitive level every reporter and every editor realizes this. It is less common, and less comfortable, to think of the negative effects of fear being news.

Media people must also worry that they are going to say something is okay, and then the *what if* disaster is going to happen. The bad thing, the feared *what if*, from time to time, will happen, and nobody wants to be in the position of having said the danger is exaggerated only to have disaster strike. Even if 99 times out of 100 the reassurance proves to have been correct, that one time reassurance is followed by disaster will hurt, and the hurt will last for longer than the effect of the other 99 times. Remember, the pessimist is never disappointed. "After all," he can always say, "Disaster almost did happen!"

That in itself is a phobic train of thought. It is hard for media people, who feel themselves in a position of responsibility, to tell people that a threat may be exaggerated. Because there is interest in disaster, and because the reporters are worried that if the worst case should occur and they do not signal it in advance, they have not given the facts to the people. They would be found guilty of minimizing a potential problem, a *what if*. They would be "in bed" with the "authorities," and they do not want to be found guilty of that. So they must talk about the risks and be cautious of reassurance. For example, at Three Mile Island the bottom line in the media coverage was not that no one was hurt. The bottom line was: we narrowly averted a nuclear holocaust, a nuclear catastrophe, the ultimate. As Walter Cronkite said, "The world has never known a day quite like today. It faced. . .the worst nuclear power plant accident of the atomic age." This is the bottom line of the public's awareness. The Three Mile Island experience is going to be a dominant image, probably for the next decade, because it has been burned into the collective national consciousness largely through television news.

One concern repeatedly voiced on the TV news tapes was that of "bias"—the concern that the "experts" were "in bed" with the industry. This goes to the heart of one of the best and worst qualities in this country. Right from its beginning, we have had an anti-authoritarian bias. We have had the shared conviction that people in charge, whether George III, or the President, or Chairman of the Board of General Motors, the big guys up there, are feathering their own nests at our expense. This appraisal has often proved both correct and helpful. The problem with it is that anybody who has any expertise or any authority is defined as antithetical to your and my personal interest. The most curious aspect of this process is the "they." That pronoun is often used in these tapes. "They" say this, or "they" say that. The "they" usually refers to authority figures. And the presumption is that "they" are not people; that "they" do not want to do right; "they" want to poison kids, destroy communities, blow up factories. "They" are also conspiring together, "in bed," against "us," or so it seems to be implied.

"They"—the experts—have been tainted by their power and their knowledge. It is magical thinking. "They," by their expertise, have become alien from you and me. It is now hard to view the nuclear issue, and in fact, all of energy is increasingly like this, without this attitude. It is hard for the average person to relate to the technical issues in nuclear power. "They" are so remote that it is almost as if anybody who understands the issue of nuclear power cannot be like you or me. He cannot be credible right from the beginning. You almost feel that he has crossed over some invisible line, that he has become a "they." And you cannot trust him.

On the other hand, there is the person who says, "There is a cover-up. You have not been told the whole truth. You are not getting the straight facts. It really is even more dangerous than you think." That person has instant credibility. I think that our society is more vulnerable than most others in the world to such criticism. The most masterful statement on these tapes was Ralph Nader's judgment to the effect that, "All the public needs to know is that there is radiation in the nuclear power plant and it's a thousand times more radioactive than the blast at Hiroshima!"

We talk about mass phobia or irrational or exaggerated fears. Americans especially are vulnerable to shared fears because, first of all, the impact of television is surely greater in the United States than anywhere else in the world. Also, we have a longstanding tradition of anti-authoritarianism that has registered an incredible increase during the last decade. We are less protected than most other cultures from our own inner fears because of our inability to trust people in authority—especially in nuclear power.

Now if Ralph Nader's opinion is the last word on nuclear power, then the answer is, "We have got to get rid of it." And that is his point, as it appears on these tapes. But it seems to me that one would want to know more about the risks and benefits of nuclear power than that.

Following the Nader concept through, for example, let us take the automobile. If "all you have to know" about the automobile is that you can get killed by it, then you do away with the automobile. Or, if all you have to know about the airplane is that it can crash and kill you and that once you are on that plane there is nothing you can do to protect yourself, you would not go up in the air. If all you had to know about food is that you can get fat and have diabetes and heart attacks, you would, I suppose, do away with food. This is a strange way to look at any issue. It is what psychiatrists call "Primary Process" thinking, which is characteristic of dreams and psychotic thought. It is also characteristic of much of our unconscious reaction to television and oral communications. This thought mode also interconnects with nuclear power in another important way: primary process thinking is, by definition, all-or-nothing thinking. Thus in the nuclear power controversy complex ideas and opinions often get reduced to simple "pro-nuke" or "anti-nuke" stereotypes. While it is easy to deplore this tendency it is more realistic to understand it as a necessary part of much of our communications process both through the media and person-to-person.

You can understand how absurd this process can become if you think about water instead of nuclear energy. I can see the television picture closing in on a swimming pool, and the voice-over saying, "Do you realize that there is enough water in this pool to drown 100,000 people?" It is true; and it is the kind of statement we often hear today about nuclear energy. All these statements magnify our fears and remove meaningful perspective. This is a powerful

34

technique of persuasion related to nightmares. Nightmare thinking includes, usually, the sense of the thinker not being able to do anything about a disaster. You are powerless. You are, in a way, a spectator. But you are vunerable. When you wake up, you can do something about whatever it is that is bothering you. But you usually cannot in a nightmare, and that is the way phobic thinking develops. It is an involuntary process of blooming, buzzing, *what ifs*. The phobic terror has a life of its own, like nightmares. The Three Mile Island crisis was like a nightmare because it was largely unreal: the disaster did not physically happen. Calling Three Mile Island a nightmare captures the concept that it is something in your mind, that the terror is in you. But with Three Mile Island the terror was shared by millions of people, and that made it that much more difficult to cope with. That is what is unique about mass phobic fear: sharing reinforces the fear and perpetuates it. The fear is authenticated by being shared.

There is a history of shared phobias. In the 16th Century, following the discovery of the New World, syphilis, apparently picked up by the explorers, went through Europe like a plague. People developed enormous, shared fears about syphilis. The French called it the Spanish disease; the English called it the French disease. It was not clear where it was coming from and how it was happening. It also had a moral dimension. In the 19th Century, the shared health phobia was tuberculosis. It became the symbol for risk, danger, vulnerability. Tuberculosis was the disease of the age, and it could hit anybody. You did not know who was going to get it. In the 20th Century America, our health preoccupation is fear of cancer.

Some of the t-shirts that came out of Three Mile Island said: "I survived Three Mile Island." Those people survived fear and terror. They survived the *what ifs;* they survived the demons in their own minds, and those demons are the most frightening problem we can face. They are scarrier than anything we have in real life, because we do not have the capacity to control our own nightmares.

Nuclear energy does not lend itself to good pictures, and the logos that the networks used for nuclear stories were often negative. One such logo is the sign of the danger of radioactivity, which is a fascinating choice because it is clearly negative; a warning of danger. Another logo used on CBS was the hexagonal sign with an atom symbolized inside. While it lacks two sides of being the common octogonal "stop sign," it is so close in form that it is highly suggestive of the real thing. It implies that what we should do about nuclear energy is to "STOP" it. Using that logo is like doing a show about medicines using the skull and crossbones as background.

Another logo that was used was the atom with the circling electrons around the nucleus, another was a plant or factory. Those were better, more neutral. A recent logo the networks have moved to is a stylized nuclear plant. There is nothing very distinctive about a nuclear plant, other than those big wide cooling towers. Sometimes the networks used a picture of an atom bomb going off, or they used the footage of what is inside the nuclear plant. They showed, for example, the mechanical hand setting a rod in the reactor core. That has a negative connotation. It is alien and strange. Seeing the plants in a more human context is preferable. For example, it would be better to see the workers at the plant rather than the mechanical hand. When I viewed the tapes, I was reassured by the sight of the actual workers at Three Mile Island simply because they were obviously ordinary human beings. One of them was interviewed at the change of shifts. He appeared to be an honest, hard-working guy doing his job.

Although I heard several times on the tapes that nuclear reactors cannot explode, it did not get across clearly enough. The other thing that did not get across was the practical meaning of the "dreaded meltdown." What does "meltdown" really mean? What would happen if meltdown occurred? I heard sloppy descriptions about what that meant, but nowhere did I see an explanation of what the actual risk was, and how great it was, or what the outcome would be after meltdown. Throughout the tapes, the fear was usually only partly articulated—hinted at darkly. I had a feeling that nuclear energy was alien to many newsmen and that they ought to have known more about it. It is important that media people be better educated about nuclear energy, or at least to have them have access to, and use, better educated advisors.

They might then run the risk of being called "in bed" with the industry. They might lose some of their "credibility" if they knew more about nuclear energy. Anybody who says anything that is not reinforcing the scare is now questioned. From that point of view, the fact that President Carter went to Three Mile Island was important. He put himself in the middle, personally, with his wife, at the site during the crisis. That was the kind of reassurance we need a lot more of. Reading about Three Mile Island and seeing it on television, you got the idea it was a very dangerous remote place to be until the President and Mrs. Carter arrived. That undercut the scare image. The President's visit ironically occurred at the same time they were evacuating the pregnant women and the children from the surrounding area, and the fact that he would go there told you something important about the levels of radiation.

Fearfulness is not a universal reaction to nuclear energy. As I watched the tapes, I was struck by the tremendous resilience most people have. The capacity of the public to endure risks and accept them is much greater than the media presentation would lead us to believe. There is something about the way the network producers select what we see which appears to exaggerate the risks and exaggerates the public opposition to nuclear power.

A lot of that resistance to fear on the part of the public has to do with ordinary denial. Many of us get through a lot of tight scrapes in life by simply not looking at the risks or saying, "so what?" and just going ahead. This fearlessness is not necessarily positive. These same characteristics let people, for example, ride a motorcycle, ride in a car without seat belts, smoke cigarettes and do a lot of other risky things. Within any population there are people who are more fearful and others who are less fearful. In retrospect you can say that fearlessness is more or less appropriate in different situations. I believe there is a lot more fearlessness out there about nuclear power than the media, in these tapes at least, leads a viewer to believe. The people who covered the news during the crisis at Three Mile Island often seemed to be surprised by that. They found, for example, in the middle of the crisis, people who would not leave their homes near the plant. I remember one woman who said, in effect, "I'm not going to leave my home when they still have people on the Island. They have workers on that Island; they're letting them go there. Then why am I going to leave my home a few miles away? It doesn't make sense. If it was that dangerous, they wouldn't let people on the Island."

I saw much non-phobic thinking on the tapes. There was a lot of sensible dealing with the immediate realities of day-to-day activities. I was struck by Pennsylvania Governor Thornburgh saying that he was impressed with the courage of the people of his state, who were going about living their lives. People have a tremendous capacity to deny risks.

My own expectation of how the nuclear phobia will play out is that we will, as a nation, reach a point of boredom. Most phobic people get over their phobia by becoming bored stiff. Boredom overcomes fear if the experience is repeated often enough and frequently enough. The idea of going over a "terrifying" bridge another time is just of no concern; successful patients in an anti-phobia program have gone over the bridge so many times, that they overcome their fear of it. In overcoming any phobia the person must get into the phobic situation for this to occur. If he avoids the situation, the phobia persists.

The nuclear risk issue itself will not be resolved. The public will never resolve it, anymore than it has resolved the issue of airline safety. There is still a risk; it is still unpleasant to contemplate. So are coal mining and thunderstorms risky. We really do not resolve these issues of risk; we just get bored with them. It is like the 50,000 annual highway deaths. We are bored with that statistic. Or the cigarette smoking deaths: there are 300,000 premature deaths in this country from smoking every year. Everybody knows that, and we are bored silly by its restatement. That is not news. We have no deaths from nuclear power and it is a number one news item, while the 300,000 excess deaths from cigarette smoking is a non-item.

One of the major reasons for this paradox is that we are bored with hearing about the actual health hazards of smoking, while we are terrified by the potential hazards of nuclear power. I think we will, one day, get bored with the nuclear fear too.

The contrast between fears of cigarette smoking and fears of nuclear power leads to another observation. The known and familiar, if devastating, health consequences of smoking are experienced by most smokers as resulting from their personal choice. By contrast the risks of nuclear power are risks an individual does not choose. They are imposed by remote and, for some, frightening "authorities," "politicians" and "bureaucrats." Many people are uncomfortable about being reminded that they have to change their lifestyles to improve their health. It is far easier to feel outraged about faceless bigwigs who are threatening us!

The TV news and the other media operate under a particular handicap now when it comes to educating the public about nuclear power. The issue has been defined as "controversial." Therefore, the electronic media must show "balance" or "fairness." For this reason, each time some information is put on the air which may be perceived as "pro-nuclear", TV news producers must find a contrasting point of view to show "balance." That commitment to "balance" gives a substantial opportunity to the anti-nuclear advocates since many of the purely educational aspects of the issue can be seen as "pro-nuke." The viewing public will have a harder time learning about nuclear power in this context of controversy which by its nature lends itself to the conclusion that "nobody knows what is happening or the experts would not disagree the way they do." On the other hand this dialectic technique is becoming more familiar and as we learn to learn in this way the depth of our understanding becomes greater.

While fear is news, and usually news is fear, it is also true that fearlessness is far more common than fearfullness. For most people fear is a temporary short-term emotion. Thus there is a biological and psychological "exhaustion" factor which will tend to favor the pro-nuclear side of the argument. In fact, the more the issue of fear of nuclear power stays at the front of our concerns the quicker and more certain is the fall of the fear. From my work with phobias I have learned that avoidance of the feared situation is, usually, a necessary condition for the maintenance of the fear reaction.

We must accept that it is impossible to establish a standard of a zero risk for nuclear power. When President Carter appointed the commission to investigate the Three Mile Island incident, I believe he said we have to prevent this type of accident from "ever happening again." Now if that is the standard, it is hopeless. Such a goal is analogous to saying, following the recent American Airlines DC-10 crash in Chicago, that we are going to prevent all further airplane crashes. That is an unrealistic goal. The only way to prevent further airplane crashes is to stop flying, and the only way to prevent nuclear accidents is to stop producing nuclear energy.

We can reduce the public fear of nuclear power in several ways. Reassurance can help. Of course, reassurance—to be credible—must be realistic. The fact that President Carter and his wife visited Three Mile Island told me something about what really was happening there, as opposed to *what if*. And that I found reassuring.

Because of the devaluation of the nuclear establishment in the public mind, it will be essential in the next few years that much of this reasonable reassurance come from independent institutions with independent credibility. Major newspapers' editorials, foundations, health and energy organizations and even international organizations should play a role in educating the public about the facts of nuclear power. That means, among other goals, putting the benefits and risks of nuclear power into realistic perspective.

We can also reduce our fear by increasing our understanding of nuclear power and of nuclear plants in particular. More reporters and politicians and other ordinary citizens should visit nuclear power plants and learn more about them. The industry should take the initiative in establishing these contacts now, when there is no acute crisis, in order to help build a firmer foundation of understanding and personal experience with the every day, hum-drum aspects of nuclear power. Being able to label the elements of nuclear fear as unrealistic and distinguishing between the *what if's* and the *what is's* of nuclear power will help, too.

Finally, radiation exposure—the most feared aspect of nuclear power generation—is uniquely measurable. Although it cannot be seen, heard, felt, tasted, or smelled, it can be easily measured. It would help many who now fear radiation from nuclear plants to have access to cheap dosimeters which they could use to measure their personal radiation exposure. While many people would have no interest in such a device—even if it were free—some people would use it and everyone would be reassured to know he or she could use one if he chose to. This would greatly increase the sense of personal knowledge and control. The lack of personal steps one could take to cope with fear now greatly contributes to nuclear fears.

Two recent developments could serve as useful models. The widespread practice of food labeling now provides concerned consumers with usable information about the nutritional character-istics of packaged foods. In the pharmaceutical area, the Food and Drug Administration is beginning to use "Patient-Package Inserts" providing the patient directly with information about the drugs he is taking. While in both areas there are those who contend, not entirely wrongly, that most consumers do not care about this information and many of those who do cannot use the information in the form it is being presented, these two efforts represent pioneering efforts to provide information directly to those affected, information which responds to the deeply-held concerns expressed by many about food and medicines. Similar efforts should be developed in the nuclear area.

Social progress can result from feared reactions, which means that manipulation by fear is not necessarily bad. Fear is an effective tool in education. I earlier mentioned Ralph Nader. He, more than anyone else, helped us as a nation to learn to "fear automobiles" and to focus on the staggering number of preventable deaths and injuries from driving. Similarly, many people and institutions are devoted to encouraging fear, and reducing denial of danger, of smoking cigarettes. To be effective, and the resulting fear of auto accidents and cigarette smoking have meant substantial savings of lives in the last decade, there must be constructive steps which can be taken to reduce risks. Some of these steps can be taken by individuals—wear seat belts; stop smoking, or smoke lower tar and nicotine cigarettes—while others can be taken only by institutions—design safer cars and highways; eliminate TV advertising for cigarettes. The challenge today to the nuclear power industry is to find ways to channel these widespread nuclear fears in constructive ways, both for institutions and for individuals. The American Cancer Society has done a wonderful job of channeling public fear of cancer into support for research and treatment for cancer victims. More appropriate to our current discussion, the Cancer Society has

40

popularized the "Seven Danger Signals" of cancer and encouraged self-care. They reinforce medical care and give the fearful person something he can personally do to help reduce his risk of cancer. Thus, promoting fear can be useful in solving problems. Denial of problems is not always helpful: the woman who finds a lump in her breast and denies it by telling herself "there is nothing wrong" is not helping herself, even though in the short run she may reduce her fear.

In summary then, the widespread fear of nuclear energy will eventually die out. But meanwhile nuclear power advocates must recognize that for any mass fear to persist, it must strike a deep, responsive chord in the public consciousness. To deny this reality and treat nuclear power only as a complex technological problem will simply increase fear and further erode the public's confidence in nuclear experts. My point is simply that, if nuclear power is to play a meaningful role in our energy future, this fear must be accepted and dealth with directly and constructively.

III

THE AMERICAN ACCEPTANCE OF NUCLEAR POWER

William Barrett

My answer is yes. The American public will accept the use of nuclear power because, in the first place, it will have to, simply out of necessity if we are to survive on a material level that is at all tolerable. But I also believe its response will be more positive than mere passive acceptance: it will want to push forward the development of nuclear energy to its fullest capacities. The American people have always been adventurous in spirit. We did not stop our settlements of this continent at the eastern seaboard, but pushed on to wrest a country out of the wilderness. We were from the start, both in fact and in conscious principle, a people open to the future. I do not believe that in 200 years we have become so old, tired, and decadent that we would turn away timidly from the challenges that the future of civilization now presses upon us. Only a failure of nerve on a colossal scale, which would be equivalent to our death wish as a great power, could lead the American people to a renunciation of the possibilities of nuclear technology.

I state these conclusions at the outset in order that there may be no doubt where the reflections in the following essay lead. Accordingly, we may now proceed to the essay proper.

The popular reactions against nuclear technology exhibit a tangle of motives that are not always easy to sort out. The public response to the shut-down at Three Mile Island is a case in point. Now that the commotion has subsided, we are able to take a longer and more sober view of the whole incident. For one thing, the reporting of what happened, and its potentialities for destruction, were vastly overblown. To be news, really news, a story should be exciting; and if it is not exciting, the journalist usually is driven to pump up its possible excitement. In this case the media came through: reporting of the whole affair rang with menacing and melodramatic overtones, as if the aim were to generate a mass hysteria. Yet for anyone who sifted through the public discussion looking for some intellectual grasp of issues, the conclusion seemed unavoidable that we as a nation are not clear about our own technology or the problem of technology generally. And unless we get clear on these fundamental matters we should not be able to sift the tangle of responses and attitudes that now surround the question of nuclear power.

Our procedure in the following, therefore, will be to consider in *Part One* the arguments for and against technology in their fundamental and stark simplicity. By taking these arguments in their simple structure, one strips the matter of encumbrances, so that one can get down to the basic human issue and the kinds of facts that are involved. One can then proceed to introduce qualifications in the arguments as these bear on the question of nuclear power.

But the logical arguments are not fully intelligible apart from the historical background of the whole question. In *Part Two,* accordingly, we shall try to deal with this historical background, not journalistically but philosophically, in order to make clear how technology enters essentially into science and thus into the making of the modern mind since the 17th century. To seek to abandon technology, or otherwise to reduce its possibilities, is to relapse to a historically lower level of consciousness. And here again—this time for inner and long-range historical reasons—it becomes clear why nuclear energy should be in the forefront of the whole technical question.

Finally, in *Part III,* we may seem for a while to go over to the anti-technological party—though not really. To deal adequately with an opponent one has to be able to enter into his point of view; and the fact is that technology, like everything human, has the Janus face of good and bad, that it does bring certain undesirable consequences in its wake. But it becomes necessary to sort out carefully what these problematic consequences are and how they can be met. What we shall find is that in most cases the anti-technological protest is motivated by the kinds of longings that, though perfectly justifiable in themselves, are here misplaced. The youth who takes part in an anti-nuclear demonstration feels that he is accomplishing something spiritual thereby; and it would be desirable if he could be shown why he is or is not.

44

I. Technology, Nuclear Power, and their Place in Civilization

It is not difficult to understand the human feelings that lie behind the anti-technological attitude, and even to enter into them. Indeed, we all share them to some degree or other insofar as we are living in the age of atomic weapons. In the background of every sensitive mind, nearly or remotely, there looms the spectre of atomic warfare and the catastrophic destruction it might bring with it. The word "nuclear" begins to have a frightening resonance, and for some minds there is no clear-cut separation between nuclear bombs and nuclear reactors: they both participate in some dark and malign power. The youth who sports the slogan "no nukes" has probably not separated out in his mind whether the "nukes" he intends are bombs for warfare or reactors for the peaceful provision of energy: for him they are all part and parcel of the same diabolical package.

Moreover, this unconscious confusion which is pervasive of the general attitude is not mitigated by the present world situation. Here today, at this point of history, the two super-powers, the United States and the Soviet Union, and both with nuclear weapons, confront each other in a jostling for power that so far no amount of negotiation or accommodation has been able to resolve. And, in the fact, the present course of events is such that the rivalry has become intensified as it begins to cover more areas of the globe, and matters seem hastening toward some final confrontation. And before that awful prospect it is only natural that we should feel a certain amount of fear.

What, then, is to be done? The choices we make will be in good part forced options—choices that we would not want to make if conditions were ideal. It would have been well perhaps if atomic weapons had not been invented before some peaceful rule of law had become firmly established for the whole earth. But conditions are never ideal, and atomic weapons are a fact in our existential situation with which we, as existing human beings, have to deal. In one sense, there is nothing new in this: our choices, even in daily life, are more often than not forced options. But usually, and this is the novelty of the present situation, they are not forced upon us so threateningly, nor does the possibility of such vast destruction loom in the background. And before this possibility it is only natural that some of us should panic, curse technology altogether, and seek to renounce it as the work of the devil.

And here I think we arrive at the real genesis of the anti-technological attitude that has become prevalent in some quarters today. Moreover, we have here a simple and clear-cut position, which in its very simplicity permits us a chance to lay bare the basic issues behind the whole question we face. (As I said earlier, we shall then proceed to the complications that have to be dealt with, but these complications, we shall find, do not remove the force of the basic principles involved.)

The position, indeed, is appealing in its very simplicity, and there is perhaps in all of us some idyllic portion of our being that might be tempted to subscribe to it. But though we may be tempted, there are three compelling arguments—two of practical urgency, and one purely moral—that I wish to advance why we cannot accept this renunciation.

(1) First: if we were to renounce our technology, the other side is not going to follow suit. The purists who would like to rid the earth of technology would not have accomplished their purpose. The circle of technology would in fact expand, carried on by a power that brooks no criticism from its citizens, and we would be enclosed within that circle as a captive people, deprived of our liberties, fabricating spare parts for the Russians.

This paper is not directly political in its intent; we speak as a philosopher and not as a political commentator. And yet some notice, however brief, must be taken of political realities, for they are part of the total picture with which any philosophical reflection has to deal. The fact is that the survival of freedom in the world today depends on the continued presence and power of the United States on the international scene. Were we to sink to the level of a second-rate power, Western Europe would quickly slide into the Soviet orbit; and Japan would be left as an isolated island that could not exist long in its present form against the solicitations and pressures of the Communist states. Hitherto in this contest the United States has enjoyed the advantage of a more advanced, copious, and varied technology with the immensely greater productivity that technology allows. It is, sadly, almost the only advantage we enjoy. The ideological advantage is all on the other side: the Soviet Union speaks with the ideology of the masses, with the inherited slogans on behalf of the poor and oppressed, and in the so-called Third World—which is really four or five different worlds, though united in their relative poverty—the impoverished masses are the over-riding social reality. This, indeed, is one of the most ghastly ironies of the present situation: that one of the worst tyrannies in history should now wear the mantle of champion of the masses and of "progressive" policies. For the impoverished masses of the Third World the truth about a Soviet system—that it is politically oppressive and that its economic performance would not solve their economic problems—is too remote, subtle, and intellectual a matter to be grasped or believed. What they have before their eyes is their own poverty, and the galling presence of those few in the society who have more and in the expectation of which they cannot hope to share under their existing society, and they are ready to listen to the utopian slogans. The United States has no such utopian idology to offer, and under the distortions of local propagandists can be made to appear as a "reactionary" ogre.

For the crusaders against technology these political realities seem to weigh little. They seem unable to realize that what stands between their freedom—including the freedom to be dissenters—and the totalitarianism that would obliterate that freedom is the continuing power of the United States, which has to include technological power. With some of the more crusading anti-technologists the political motive is fairly clear: they are partisan to the ideology of the opposing side, and they generally have an adversary relation to the United States. They do not seem to mind if the United States, with its present institutions, were to disappear. Others in the crusade are less focussed politically and more simply utopian. They are not sensitive to the reality of power, and the balance of power, in the present international situation, because they have a distaste for power in any political form and do not want to think of it. They are the heirs of a certain extreme kind of Liberalism which affects to think of the Social Contract apart from any involvements with the question of power. Power is a dirty business, they think; and they are probably right. But dirty or not, it is an indispensable aspect of any social organization. Even a utopian like Plato had to develop a military class within his Ideal Republic to protect it from threatening neighbor states. In our present world the United States can survive as a free country only if it is buttressed by a highly developed technology and by technical armament.

(2) The second argument I have to offer might be called "The Rubicon Argument." You will recall that when Julius Caesar crossed the Rubicon, a small river in northern Italy, in his march on Rome, he made the famous declaration, "The die is cast"—he had nowhere to go but forward. He had crossed a frontier laid down for him by an edict of the Roman Senate, and now he could only push forward or perish. In the matter of technology mankind has long since crossed its Rubicon—had we ever made the decision to renounce technology, we should have done so long ago, perhaps at the beginning of the 19th century—after which the population curve has climbed steadily and dizzily upward as our technical capacities could support more life. Were we to renounce technology now, millions of people would

47

perish and many more lose any hope of ever rising to a satisfactory standard of living.

Anyone who has looked at these population curves for the Western world since the year 1800 must be struck by them as awesome. They present us with an overwhelming fact about the modern age that many social thinkers are prone to forget. The philosopher Ortega y Gasset noticed this in the 1920's in his *Revolt of the Masses* and emphasized it as a key point in our attempt to grasp modern history: *"There simply are more people."* Only in the last two decades has serious thought been given to the possible dangers of overpopulation, and various projects and plans have been suggested for limiting the population growth. But even if one such scheme should work (which is not at all certain) and the population of the planet should become stationary at about its present level, the problem of supporting that population would still be formidable. We simply do not have enough operant technology today to maintain all the people of the globe at any decent human level. Merely to divide what we have would only produce a general scarcity; we need greatly added production, and for that we need more technology. And when further, we take into account the dwindling of natural and irreplaceable fuels, we shall be compelled to create all kinds of new technologies to fill the gap. Technology is not a static collection of apparatus and hardware but an unceasing process (as we shall see in the next Section).

Thus, simply from the point of view of material subsistance, mankind has already crossed its Rubicon on the question of technology. We are bound to technology. It has become a destiny mankind cannot now escape even if it wanted to. But to accept a destiny is not to submit passively to an inert load; it is to *enter into* the destiny; we have to accept and shape it.

(3) The third, and final, of our arguments takes us beyond the realm of the practical exigencies, either the fearful exigencies of the present international situation or those of longer run that have to do with the material subsistance of the species. It is a purely moral argument that is this: we are under the simple moral obligation to develop our technical capacity because the power to create technology is a uniquely human gift and we are under moral obligation to develop our gifts if we possibly can. We recognize this obligation clearly and distinctly in the case of the individual. When a young man or woman has an unusual natural talent—for mathematics or music, say—and allows it to languish, we declare this neglect to be a regrettable waste. I believe we must make a similar judgment for the species; since humankind has been endowed with the gifts—

God-given or nature-given, as you choose—of intelligence and manual dexterity to fabricate a technology, it would be morally wrong not to develop these gifts as much as we can. The obligation is all the more compelling when we consider (as we shall in the next Section) the intricate and internal relations of technology and modern science, such that to restrict any particular technology may very well slam the door upon some form of human knowledge. Man is the creature who seeks to know; and to frustrate this natural desire is a sin against our human nature. Needless to say, we are also morally obligated to develop this technical capacity with all possible precautions for human safety and human values.

These arguments have been broad in scope, and simple in structure, and deliberately so, as we have already indicated. And now we must proceed to the complications and qualifications that must be added to them insofar as they bear particularly on the technology of nuclear energy. Nevertheless, whatever these qualifications, the simple force of these arguments will hold in each case.

And indeed, one can imagine the crusader against nuclear energy already snorting against the simplicity of our argument. "I am not," he will declare, "against technology as such. What a misrepresentation of my views." We are here, let us take it, dealing with the more sophisticated spokesmen of the anti-nuclear crusade. As for the rank-and-file of the crusaders, it is a very arguable point how far their basic attitudes are permeated by an anti-technology bias, for many of them still have a nostalgic yearning for the counter-culture of the 1960's. But we will not press the point here. "What I am against," we may imagine this suggestion continuing, "is not technology, but nuclear technology particularly. And I am not even against that technology *per se*, but against the present operation of nuclear plants until they have been shown to be completely safe. The dangers of nuclear radiation are too serious for us to trifle with."

This looks to be a much more limited condemnation, but its limitations are more apparent than real. A certain spirit of tolerance, it might seem, is shown in the willingness to accept nuclear technology once it has been shown to be safe. But what is the margin of safety that is demanded for acceptance? I am not a technician, and it is neither my aim nor my competence, to go into the technical precautions for safety; but one cannot help bringing forward a simple fact of common knowledge that there have been 200 nuclear plants operating in the Free World over the last ten years without a single serious casualty. That is a record of safety unmatched by any other of the major technologies that we have. And if some greater and unspecified margin of safety is demanded, how, pray, are we going to attain it if the technology in question is to be banned meanwhile? We improve and perfect a technology only as we put it to use and refine it. Consider the technology of the steam engine (with which there were many more casualties in its early development than have ever attended nuclear energy), and particularly its use on the steamship. How would we ever have known if a steamship were capable of crossing the ocean if we had never attempted the voyage? We would still be sitting on our hands waiting—and waiting for what? For someone else to make the voyage we were too timorous to make ourselves. (And that, as we shall see in a moment, is the crux of the problem insofar as it concerns the level of our technical capacity and its possible decline.) The point is that the voyage *was* made, and thereafter steamships became safer through constant practice and traffic. And so with other technologies, as, for example, with the airplane.

And indeed, the limitation of the crusaders' ban to nuclear energy plants cannot logically stop there but must extend bit by bit to ever larger areas of technology. And this can be seen if we turn back, very briefly, to our three arguments above. Consider the first argument (1) which had to do with the American position as a world power in the present international situation. If one is crusading against nuclear plants engaged in the production of energy for peaceful purposes, why in the name of logic should one's protest stop there? Why not seek a ban on atomic weapons generally? Even in the mere stockpiling of such weapons we cannot be absolutely certain—we can, alas, be absolutely certain of so very little—that some careless mishap might not release an atomic conflagration and pour the dangerous radiations abroad. We would be logically led to a ban on atomic weapons altogether; and if we could not persuade the other side to join us in it, which is altogether likely, we would be obliged, in the moral rigor of our position, to follow the course of complete and unilateral disarmament. This is a possible position among the uncomfortable options thrust upon us by the present world situation; it is not a position to which I personally subscribe, but it is not the issue in the context of this discussion and I do not propose to debate it here; I would like only to point out those of its undesirable consequences which go directly against the anti-nuclear yearnings of the crusaders. For one thing, our unilateral disarmament would in fact be a surrender to the other side, and would bring with it the grim and oppressive society that the Soviets carry with them wherever their power spreads. And, in that case, what would prevent them from enjoining upon us as a captive nation those experiments in nuclear technology which might be truly dangerous, since it would be our land and population that would run the risk in place of theirs? Thus our first argument holds even when we restrict its scope particularly to nuclear energy.

Similarly, argument (2) above holds just as effectively when we restrict the scope of the ban to nuclear technology only. The anti-nuclear crusader tells us that he is not against all technology but only the technology of electricity-generating reactors until it is proven absolutely safe. But whoever has any idea at all of the extent and depth of human poverty on this planet will hesitate at a suggestion that we diminish our powers of production anywhere along the line. We are told that we should develop alternative sources (which, of course, we certainly should) like solar energy and coal. But the over-riding point here is that you cannot arbitrarily cut out one whole part of technology and say the rest is unaffected and you place no ban on it, for the fact (and this will be developed more fully in Section II) is that the whole of technology is an intricately interlocking human project, and all the more so the more advanced it becomes, and you cannot remove one part totally without seriously diminishing the potency of the whole. The crusader cannot logically restrict his ban; when he demonstrates for closing nuclear plants, he is both in logic and in fact crusading against technology as a whole.

Precisely the same point holds in the case of our argument (3) which concerned our human and moral imperative to explore and to know. The fabric of technology is interwoven with the whole fabric of science, and to foreclose on one part can always conceivably shut the door on some other possibility. If the crusader wishes to shut down our nuclear plants, despite their extraordinary record of safety, why should he not storm into the laboratories of our nuclear physicists, who may be doing quite daring researches into the nature of matter, and demand that those laboratories be closed pending some guarantee of an absolute safety? If the crusader has not yet done that, it is because the name of science still commands a certain awe and respect among us, and because what the scientist is doing is not yet in the public domain. The next step would be to insist that it be always and everywhere in the public domain, and the scientists' credentials for every piece of research be subject to ratification by the crusaders—in which case, of course, the whole project of science as free and unhampered inquiry in pursuit of truth would be most seriously jeopardized. One can imagine, indeed, had "the movement" existed in those days, demonstrators marching past that abandoned squash court in the University of Chicago where in the early 1940's Enrico Fermi, fashioning the first controlled atomic reaction, virtually launched the Atomic Age. Was what Fermi was doing absolutely safe? Since it was a pioneer reactor, how could he be absolutely sure that it would not blow up in his face and contaminate the whole environment? (In the nature of the case, since it was a first trial, Fermi could not have had the safeguards that a modern nuclear reactor enjoys.) The Atomic Age might never have been born.

We need not tire the reader here by multiplying details to show that the force of our original arguments holds even if we try to restrict the ban to nuclear technology exclusively. One very central point, which we have not brought forward before this but which will emerge in the following section, is that nuclear technology must be regarded as the most advanced part of our technology simply because it springs from that part of our science which represents man's deepest and most daring step in the knowledge of matter, and therefore of the material universe that surrounds us. And to block this advance tip of the wave is to impede the whole movement of the technological current.

II: The Historical Background; Technology-Science as the Decisive Factor in the Modern Epoch

What is technology fundamentally, and what is the nature of the technical order under which we are coming more and more to live? These questions cannot be answered adequately unless we place the whole phenomenon of technology within its historical setting. Historical interpretation here does not merely add some incidental facts that have a casual or merely illustrative relation to the logical analysis of the matter; on the contrary, it is only when we place formal analysis within its fullest and most concrete historical context that the real sense of the former takes on its real meaning. The history and logic of the technical phenomenon are in constant interplay.

In the first place, we do not really understand technology unless we have grasped its peculiar and intrinsic relation to Modern Science. Many people, perhaps most people, think of technology merely as the practical application of science at some points where the conclusions of science make contact with practical life. Science, that is, goes on its way independently of technology, and by some happy or unhappy accident a technical application may be hit upon. Such approximate accounts, of course, are not altogether false, but they do not bring us to the heart of the thinking that created modern science in the first place and is still operant within it.

The Seventeenth Century, which created Modern Science and which Whitehead has aptly called "the Century of Genius," was not unaware of the revolution it was introducing. Men like Galileo—who called it "the New Science" even while he was in the process of launching it—were militantly conscious that it was a break with the past, that it introduced a way of thinking unknown to the Ancients, however wise and subtle they had been, indeed that it was what we might call today a new mutation of the human mind. For the sake of brevity we may confine ourselves here to a single thinker who was in a uniquely favored position to grasp the revolution that had been wrought. Immanuel Kant, the greatest philosopher of the modern period, came a century after the great founders of the New Science. He had been a diligent apprentice of this science in his youth, and continued unceasingly thereafter to reflect upon its meanings. Moreover, where great forerunners like Descartes and Galileo were talking about something that existed largely as their own *project*, by Kant's time this New Science has come fully into existence and had already given a stable and awesome body of knowledge to the world. We may take Kant, then, in a single quotation from the Preface to the Second Edition of his *Critique of Pure Reason*, which— for those intrigued by the symmetry of dates—appeared in 1787, just one-hundred years after the publication of Newton's *Principia* in 1687:

> When Galileo caused balls, the weights of which he had himself previously determined, to roll down an inclined plane; when Torricelli made the air carry a weight which he had calculated beforehand to be equal to a definite column of water . . . a light broke upon all students of nature. They learned that reason has insight only into that which it produces after a plan of its own, and that it must not allow itself to be kept, as it were, in nature's leading-strings, but must itself show the way with principles of judgment based upon fixed laws, constraining nature to give answer to questions of reason's own determining . . . Reason, holding in one hand its principles, according to which alone concordant appearances can be admitted as equivalent to laws, and in the other hand the experiment which it has devised in conformity with these principles, must approach nature in order to be taught by it. It must not, however, do so in the character of a pupil who listens to everything that the teacher chooses to say, but of an appointed judge who compels the witness to answer questions which he has himself formulated. Even physics, therefore, owes the beneficent revolution in its point of view entirely to the happy thought that while reason must seek in nature, not fictitiously ascribe to it, whatever is not being knowable through reasons' own resources has to be learnt, if learnt at all, only from nature, it must adopt as its guide, in so seeking,

that which it has itself put into nature. It is thus that the study of nature has entered on the secure path of a science after having for so many centuries been nothing but a process of merely random groping.

What Kant is pointing to here is an intrinsic relation between science and technology. The scientist's mind is not a passive mirror that reflects the facts as they are in themselves (whatever that might mean); the scientist constructs models, which are not found among the things given him in his experience, and proceeds to impose those models upon nature. And he must often construct those models conceptually before they are translated at any point into the material constructions of his apparatus in the laboratory. In this connection, indeed, Kant could have used a simpler and a more radical example from Galileo than the famous experiment of the inclined plane, and that is Galileo's very construction of the concept of inertia itself. That concept had hitherto been lacking to the investigators of nature, and the science of mechanics (which was in fact basic to the whole of the New Science) could not get under way without it. What does Galileo do? He does not record passively the facts around him; instead, he constructs a concept that is not precisely found in nature, and indeed is contrary in certain respects to what we find there. Imagine, he tells us, a body on a perfectly frictionless plane; if motion is imparted to this body, it will move on infinitely in a straight line unless its course is impeded or altered by some countervailing force. Well, nature does not present us with any perfectly frictionless planes, nor with any plane that is actually infinite in extension. What the concept constructs is a model that is actually contrafactual in the light of our ordinary experience. No matter; it serves as an ideal standard in approximation to which actual situations may be effectively calculated. Here the basic concept of the science, since it is entirely man-made and does not literally copy any single fact in nature, is a product of human artifice and therefore a technical construct as fully as any material apparatus. Here science is technological at its very source, in the formation of basic concepts, and not subsequently and as if by happenstance in the practical applications it may find. The hyphen in the compound expression "science-technology" does not signify the compounding of two independent entities only externally related; it expresses a single historical reality of which the two names denote merely different aspects.

And this intimate association of the technological and the scientific, which indeed begins at the root, spreads through every branch and leaf of the whole tree. The more advanced and developed

54

a science becomes, the closer its alliance becomes with its own technology; and the more closely interwoven part becomes with part in the unity of the whole. One cannot say, antecedently, that any particular fact within the scientific structure, no matter how isolated it may appear at first glance, cannot have an unexpected and surprising connection with some other parts within the whole. The surprising discoveries of such connections, where they had not been at first even suspected, have been a constant and recurring feature of science in its actual development. And so too with any particular technical device, no matter how isolated and particularized for its own small function it may appear: one cannot say of such a device with antecedent certainty that it, or the principle of its operation, will not find an indispensable use elsewhere. The moral of this, particularly in the context of our present discussion, is that you cannot arbitrarily cut off one part of your technology—especially such a crucial part as concerns energy, which fuels the whole scientific-technical enterprise—and be confident that the whole structure will go on intact and unchanged. In short, once one has grasped the unitary phenomenon that is science-technology, one realizes that it is a single human *project*, and the project moreover under which the history of the last three centuries, the history of our modern epoch, has been and is being played out.

Let us not underestimate the daring and depth of that project. It marks, as Kant notes, a turn in human reason, and, consequently, a transformation of our human being in its deepest attitudes toward the world. Humankind turned away from a passive to a more active role in its struggle with nature. Life is given us to be mastered—not as something to drift along with. Notice some of the key words Kant uses: man *imposes* his models upon nature; he *compels* nature to answer his questions; he does not merely submit, but seeks a position of *command*. These are words of power, and Kant knew very well what he was about in employing them. He had been a diligent and admiring reader of Francis Bacon, and he was very well aware of Bacon's famous dictum, "Knowledge is power." (Bacon, of course, went beyond this sensible maxim to the more violent metaphor that "We must put nature to the rack;" and this immediately raises the questionable aspects of our assumption of power over nature, (which we shall come to very shortly in Section III). Kant internalizes and deepens Bacon's point: knowledge not only is power in that it may provide us the instruments to deal with a particular situation; more profoundly still, the step toward knowledge is in its very essence a step toward power. In the very constructivity of our scientific concepts—the human mind fabricates concepts which, literally, are not found in its ordinary world—we already have taken a step *beyond* nature, in order subsequently to deal with it. Nor is there anything "unnatural" in this step; indeed, in taking that step the human mind comes into the fullness of its own nature and its powers. "God helps those who help themselves," runs the old adage; we have to use all of ourselves—including that part of ourselves which steps beyond nature—in the struggle to cope with nature for the survival of our species. Thus the project of science-technology, with which the 17th century launched our modern age, does represent, I believe, a genuine and beneficent transformation of our human being. With it, the doors of knowledge were flung open. In three short centuries we have come to know more about the nature of the world in which we live that in all the millenia of human history that preceded. And we are only at a beginning, still wrapped in the toils of ignorance. Those who would shut the doors upon those further possibilities of knowledge are the timorous in heart who begin in sentimentality but end as the stern jailers of mankind. The scientific revolution also created an openness toward the future in another area: in the expectation of social advance, progress, and the radical possibilities of improvement of our human lot upon this earth. Reformers became more hopeful, optimistic, and in some cases, alas, even utopian.

Today, to a great extent, those hopes and that optimism have retreated; and in some quarters, indeed, there prevail a certain despair, passivity, and nihilism toward the future. In our next Section, now, we must try to look at some of the reasons why this has come about.

III. The Problematics of Technology; Our Malaise at the Technical World

I have been addressing the matter at hand as a philosopher, and in this fact itself we may find some food for reflection. It is expected of me that I should speak from this point of view, for by profession I am a philosopher, and we are required to carry out the role that our profession assigns. Modern life assigns us all to our separate slots, and we tend more or less to acquire our identity, as social beings anyway, from the slot in which we have been pigeonholed. For myself I have no objection to this arrangement: I find my profession congenial, the subject of philosophy is absorbing, and for me it sheds an extraordinary light on our present civilization. But other people do not seem to be so happy about the slots into which society has cast them. They find these slots too narrow, cramping, and unrewarding. And we have here, consequently, one of the most common complaints

56

against a technological society—namely, that, in the minute division of labor which technological production inevitably entails, the human individual caught in the toils of it becomes an over-specialized fragmented being. Hence the profound *malaise* that stirs in some minds at the very idea of the technical order of modern society.

It is important to dwell on this point for a moment, for most people tend to think of technology—when they think of it at all—as primarily an affair of machines and apparatus. But just as much as an arrangement of machines and apparatus, technology represents an arrangement and ordering of the human beings who tend the machines, and the even greater number who have to tend to the various services and requirements of those products as they enter the market place. In this respect modern technology is altogether different from the technology of the ancient world because modern machines tend to create a human network around them. If you go back to the Greek philosophers, Plato and Aristotle, for a discussion of technics (that is, *techue* as they understood it) you will find that the examples they cite—a pruning-hook, axe, knife, and so on—are simple utensils that fit into the context of life as it then was. A modern machine, however, may bring with it a new human context. Consider the telephone, for example: in its inception it was a relatively small apparatus which Alexander Graham Bell could contain in two separate rooms of his home in Hartford; in its present operation, the telephone is an intricate and pyramidal organization of subscribers, service personnel, and management that make its operation effective. Without unduly stretching the point, we could say that Alexander Graham Bell did not only invent the telephone, he invented the telephone book. It is not always clear when we talk of any particular network whether we are referring to the network of apparatus or the humans who tend that apparatus in one way or another. When a newspaper speaks of a particular TV network is it referring to the web of electronic equipment, or to all the humans—management, staff, announcers, and even the audience as part of that peculiar and immense family—that are bound within that web? Clearly, to both at once, and to both as one thing. For here the fusion of mechanical and human is so complete that it has produced a new identity. And it is this fusion that some of the anti-technologists fear, for in the process, they feel, we humans may become overly mechanized.

The deep and underlying historical reality at work here is the general fact of *the rational organization of human life* as an indispensable condition of our economic and social existence. In its degenerate forms this will to organization can drift into aimless bureaucracy and mismanagement. But in its inception it was necessary toward greater productivity. It was a revolution brought in by the new capitalist class that has ultimately proved to be more far-reaching than the noisier revolutions that usually claim our attention. What lies behind this massive will to organization?

Sociologists and historians give their own accounts of the matter. Indeed, this theme of the rationalization of modern life was the principal subject—and indeed almost belongs to him—of the great sociologist Max Weber; though we have to remember that Weber, being German, had drunk deeply of the waters of philosophy. But I wish, rather, to fall back into my own slot and speak from there, thus accepting the fact of this organization even as I try to analyze it while speaking within it. I happen also to believe, with no demeaning of sociological analysis, that we are able to see the whole process more comprehensively and imaginatively, and I think therefore more deeply, from the point of view of the philosophy that lays the ground for it and thus both projects and mirrors its development.

Three things mark the revolutionary break of our modern age with the past.

(1) The first we have already noted in Section 11, and it has to do with power and control as an essential factor in science; but we have now to place it in a new light and give it a somewhat more pronounced emphasis. Kant had said we understand nature only insofar as we *impose* our models upon it and *compel* it to answer our questions. We may sum up Kant's position here in more abrupt and simple, but I think not falsifying, form as: We understand nature only insofar as we dominate it. And remember that Kant here is looking backward and taking stock of a revolution that had already been wrought. In short, what emerges with the 17th Century is a new metaphor for our human existence: the metaphor of man as the conqueror and dominator of nature.

This metaphor, moreover, takes a powerful grip on the period that follows. Consciously or unconsciously, and either in consent or opposition, the history of the modern age unfolds within the framework of this metaphor. Thus the political revolutions of modern times are launched under the aegis of this metaphor. The domination of nature includes human nature, which naturalistically considered is merely a part of nature; and the project for the future is to transform this human nature totally by social planning. The French Revolution dreams of the New Man, the *Citogen*: and the Russian Revolution of the new Soviet-man. In their aim to transform

58

humanity they are compelled to be total; they end, alas, by being totalitarian.

(2) Along with this metaphor of domination, as an unexpected consequence, something else emerges that is both disquieting and uncanny. This Nature, which is to be subdued, becomes stranger, more puzzling, more alien. Our modern period begins with a great explosion of knowledge—which, however, leaves the world more questionable than it was. The famous problem of our modern alienation, so much discussed and often foolishly discussed, has its origin here.

The so-called "Copernican Revolution" has a vital place in this. Not only did it thrust man out from his central place in the universe, thus deflating our human ego, but it opened up vistas of a universe so vast that it might exceed our human powers of comprehension. And, further, it broke with the picture of the world in traditional astronomy that presented a universe altogether congruent with our ordinary perception: what appeared to be the case—the suns, stars, and planets circling around us overhead—was really the case; the apparent world was the real world. But now, in the new theory, the real world turned out to be altogether at variance with what appeared in the perceptions of ordinary life. Thus Descartes' famous conjecture of an Evil Demon who has fashioned the universe so as systematically to deceive him is not merely an idle concoction for the purposes of dialectic. Behind it there is the anguish of a whole period (and which comes out most explicitly and eloquently in a figure like Pascal) that mankind is lost in a universe whose reasons, if any, it can never hope to comprehend. But Descartes is rescued ultimately from that uneasiness because one part of his mind remained that of a faithful medieval Catholic, and God was immediately present to him. But that presence fades as we move on to our next point:

(3) Finally, but perhaps most important, is the withdrawal of God. God does not disappear, He recedes. He is no longer at the center of human life, but—in Matthew Arnold's phrase—has retreated down the vast edges of the world. Modern civilization becomes more and more secular in all its modes, and the sense of the sacred diminishes. There is always a jolt of surprise when we return to read them to be reminded again how the minds of those revolutionary giants of the 17th Century who launched modern science—a Descartes or a Newton—were still so completely contained and satisfied within the structure of Christian theology and worship. Kant is perhaps the last systematic philosopher of Christian theism, but by his time the forces of secularization have been at work, and it may be wondered at times whether he is not embalming the residue of a religious tradition: he

59

insists on God as a moral postulate, but for the most part his critical philosophy is intent on setting limits upon the place of the Deity in human thinking. The whole effort of his philosophy, as one wit has observed and not altogether inaccurately, was to keep God in His place. And after Kant despite the brief flurry of German Idealism, which was fighting a retreating battle in any case—the whole course of modern philosophy becomes progressively more positivistic and materialistic in spirit.

But what, the reader may ask, has this summation of intellectual history to do with our present question? Well, the residues of human history are much more a part of us than most people realize, and our contemporary consciousness has become what it is through the evolution of the whole modern epoch. We are on the track in this Section III—the reader may remember from our opening—of the youth who sports a "No Nukes" button and marches in a demonstration to close a nuclear plant that is engaged only in the production of energy for peaceful purposes. We want to know what makes him tick, what ultimate motives—human and moral—draw him on, and why, though he himself may never use the word, he really thinks he is accomplishing something "spiritual" by taking part in those demonstrations.

Now, the fact is that the contemporary world, if we look at it honestly, presents us on the whole a spectacle of spiritual emptiness, uncertainty, and disarray. Three centuries of the Modern Age have produced an unparalleled advance in our knowledge and our ability to manipulate material nature, but it has also ushered in a general spiritual decline. A sense of meaningless and lack of purpose pervades a good deal of modern life. If you doubt this, you have only to look around you at the culture we produce and consume. And if you seek documentation of it, it has been done in book after book, almost *ad nauseam*. (I have done it myself in some previous works, and do not wish to repeat myself here.) Religion has ceased to be central to the lives of people, as it once was, and in consequence has declined. And nothing has come to take its place. And yet the spiritual hungers of human beings remain. Hence the need to find satisfaction of those hungers in some kind of social crusade or political cause: at least there, the individual feels, he is joined with others and is escaping from his own barren and ultimately meaningless selfishness. Even the so-called "Me" generation is now casting about for causes.

To what extent is technology responsible for this spiritual decline? As the most visible part of a fundamentally materialistic civilization, it is easy to be made into the whipping boy for our spiritual ills. In part this is a recoil from an illusion about the omnipotence of technology. People imagine that technique and technical thinking are able to resolve all human problems; and when they find out, in their own life, that this is not so, they swing to the opposite extreme: from the saviour theory of technology they go to a view of it as the devil. Technology then becomes in their eyes the source of all, or most, of the ills in the world today; and there are, unfortunately, enough bungling and clumsy uses of technology to provide them targets to attack. But the fact is, of course, that technology in itself is morally and spiritually neutral. Technology is simply an instrument that is absolutely necessary for our human survival; and, moreover, it has become necessary not only to maintain it at its present levels, but constantly to expand its powers and efficacy. The spiritual import of technology—whether it becomes a force for good or bad—depends entirely on the use we put it to. There is nothing in technology in itself that is inimical to the life of the spirit. I do not feel humanly trapped in the various appliances of my kitchen; and if I set about in a fit of rage to destroy dishwasher and refrigerator, I do not in the least raise my spiritual level. A man may pray equally devoutly on an airplane as on the solid ground; and it is an ironic fact that some people, out of panic perhaps, are often moved to pray on an airplane, but forget about that spiritual exercise as soon as they are back on *terra firma*.

What, then, is to be done in our present situation? I do not know the way to bring about a spiritual regeneration for this civilization; that falls within the province of the religious prophet, and I am only a philosopher. But philosophical reflection here may nevertheless provide us some help in correcting the ways we think about technology, and so enable us to strike a more balanced attitude that may rescue us from some of the unnecessary confusions of the moment.

For one thing, we have to strike a middle course between the saviour and the devil theory of technology. While recognizing its vast and necessary powers for maintaining our material life, we have also to become aware of the limits of technical thinking in some very crucial human areas. For this purpose we shall have in some sense to correct and balance the metaphor of man as the conqueror and dominator of nature that launched our modern epoch. That was a triumphant and exuberant metaphor in its time; but it has perhaps already served its purpose, and to carry it out further, without any qualifications or limitations, would be to pursue a course that would become demoniacal and dehumanizing. We have to learn the limits of purely technical thinking by dwelling long and carefully on those countervailing modes of experience where we must be receptive rather than dominant, obedient rather than commanding; where we can't force the issue but have to let it unfold and abide by it as it unfolds. We have to learn, in short, what parts of our experience elude technical manipulation, and where we only play havoc by trying to introduce such manipulation.

Our moral experience is an example. How do we become morally wiser? By this time we ought certainly to know that technical advance in itself does not bring the moral improvement of our species. The science fiction movies, if they taught us nothing else, show us that; there on the screen are those people of the future, with gadgets and weapons altogether beyond our present powers, and yet emotionally and morally more infantile even than we are. The charm of these movies, when they are charming, is our childish delight in spectacles: we become as little children again gazing at the fireworks of some imaginary Lunar Park. But in that mindlessness there is a foreboding of horror too: Suppose these movies foreshadow the real future in which we make spectacular progress in technology but regress morally.

There is no moral development of humanity except as the moral development of individuals. The moral maturing of the individual may have to travel a lonely path. A first step on this path is learning to be truthful to oneself. When you confront yourself in the mirror of conscience, you are not a member of a research team imposing a model upon data. If you impose anything, you are likely to impose on yourself and go on lying to yourself. The problem here is not to commandeer the truth, but to let yourself be able to receive it. To admit to the image in that mirror is hard; it is not the path of domination but of humility, sometimes even humiliation; but it is the indispensable step toward freedom. The moral life cannot be technologized; if there were a machine to make us more moral, we would simply go through the same moral struggles on another level about using the machine.

Aesthetic experience is another area beyond technical manipulation. Before the beauty of nature, or in the grip of a great work of art, we do not dominate but surrender. The highest compliment we can pay to the work of an artist is that we are enthralled by it. We would do well to think for a moment about this word "enthralled". It is a reminder our language still preserves against our modern mania for power. A thrall, in Anglo-Saxon, is a slave, a serf, a bondsman. To be enthralled is to be as a thrall, a servant, in the presence of the work. Thus the most exquisite mode of aesthetic appreciation, to be enthralled, speaks of itself in the language of humility rather than domination. What would the future be worth, then, if its technical conquistadors were incapable of being enthralled by the wonder and beauty of the nature they exploit?

Finally, we shall have to correct that metaphor of domination in the experience of awe as we face nature itself. While we must strive to use all the resources at our command in order to cope with nature and survive, we have nevertheless, face to face with the universe, to recognize our own abysmal finitude; that all our grandest achievements in technology will always take place within an encompassing reality that exceeds our powers.

Perhaps I may make the point vivid by a parable:

On my local railroad station along the Hudson River there was a spot between some criss-crossing girders where the spiders used to spin their webs. By midsummer the webs had become larger and more imposing, the spiders bigger and more ominous. They had chosen a good place, sheltered by a projecting edge of roof; and the catch in their nets—night-moths, mosquitoes, midges—was abundant.

That summer I was a constant visitor with them, particularly at night, when they gleamed malignantly under the arclight of the station. And I found myself fantasizing a spider consciousness and spider philosophers among them. A spider Descartes, who tells them: "You must spin clear and distinct webs." A spider Kant: "We must impose our webs upon nature, for only thus can we grasp it scientifically—that is, spiderishly." Or his intellectual descendant, the modern Spider-Positivist: "What is alleged to exist outside the web has no meaning, it does not exist."

Well, that year the rains came early. An unexpected hurricane lashed its tail at us, blew down the protecting edge of the station roof, and the violent rains washed out all the webs.

The moral of our tale: he who would conquer nature may end as a washout.

The spider spins his technical network out of his guts, we humans spin ours out of our brains. But each is equally a part of nature, each equally embedded in it.

So long as we hold fast to this lesson of humility, we may carry on our technology with all the prudent daring we can muster.

IV

THE NUCLEAR "GENIE": BEYOND FAUST, FATE, AND INCANTATIONS

Margaret N. Maxey

"From the Land beyond Beyond,
From the World past Hope and Fear,
I bid you, Genie, now appear."

. . .Seventh Voyage of Sinbad

The potent image of a "Nuclear Genie," presumably subject to the superior will of a powerful individual or group who can cause it either to appear or disappear by cryptic incantation, has recently been revived in a dubious form of literary coinage. Similarly, but with more negative connotations, the literary symbolism of Faust and his disreputable bargain with the devil has been frequently used to cast doubt on the moral legitimacy of continuing to develop nuclear science and technology, especially when applied to centralized production of electrical energy. The symbolic power of such images invites closer scrutiny.

I. The Problematic Status of Rhetorical Symbols

If media-generated fallout from the Three Mile Island episode and recent publication efforts are any indication, we are now entering a new phase in the rancorous dispute over nuclear technology—the phase of obituary and epitaph writing. We have been exhorted to "rebottle the nuclear genie" in Amory Lovins' blueprint for a non-nuclear future entitled, "The Soft Path: Toward a Durable Peace,"[1] and in a more recent collaborative effort to enjoin "Nuclear Power and Nuclear Bombs," selected for publication by the editors of *Foreign Affairs*.[2] We can also read the Bupp-Derian colloquy entitled, "Light Water: How the Nuclear Dream Dissolved."[3] Anti-nuclear coalitions appear close to success in their drive to eliminate any nuclear option from the energy future of the United States. The reasons they may be successful contain important signals for those who study the role of symbolic images and myths in shaping our democratic process.

On the surface, it appears that mounting public concern, criticism, and confusion over vacillating nuclear policies exist because there are still genuine unsolved technical problems with the safety of nuclear reactors, preventing the proliferation of nuclear weapons, and isolating radioactive wastes from the biosphere.

However, recurrent themes in the skillfully orchestrated political campaign against nuclear electricity suggest that the reasons for public concern derive, in fact, from non-technical origins. The strategy set forth at the organizational meeting of Critical Mass in 1974 has become increasingly successful. At that forum, Dr. Margaret Mead exhorted her audience "to make people feel that everything they value in the world is at stake" if the development of nuclear energy is allowed to continue. She stated then that "Americans aren't afraid of dying *suddenly* but of dying *slowly*." She urged her listeners to concentrate on those aspects of nuclear power that would evoke the most fear, namely, the long-range deteriorating effects of low-level radiation.[4]

Psychometric surveys over the intervening years suggest that the public has not selected physical and environmental safety as principal issues raised by nuclear energy. Rather, public concern is about the personal, social and political consequences which this technology seems to entail.[5] Psychic well-being and personal power appear threatened by the belief that radiation exposures (no matter how trivial) will have delayed, incremental, somatic and genetic effects. Social well-being seems threatened by the belief that profligate energy use will continue to escalate; developers will continue to ravage delicate ecosystems and wilderness areas; energy magnates will ruthlessly exploit non-renewable resources; rising expectations of increasing numbers of people will grow unchecked. Political welfare appears threatened by the belief that terrorists or unstable persons and governments will have undetectable access to plutonium out of which nuclear bombs might easily be made to blackmail other countries or to induce a police state or to bring down an unwanted government. In sum, nuclear science and technology have become the focus of public debate because they epitomize the major questions over which competing groups in a new political class are struggling for dominance in their bid to set a political agenda for the future of our technological society.

Public confusion about the issues raised by nuclear technology suggests also that the scientific and professional community has itself been fissioned along with the atom. The reason is clear: the divisive issues are not scientific and technical.

Why is it that nuclear fission has become a casual villian, a Trojan Horse disgorging into our midst all manner of social complaints: environmental degradation, industrial pollution, political corruption, prostitution of science, and so forth? Alvin Weinberg's celebrated lament over the "Faustian bargain" made by "we nuclear people" would not have been repeated so often with such relish and symbolic power if it had not aroused uncommonly deep passions.[6]

In the view of some people, the violent anger and premonitions of disaster associated with nuclear fission dramatize the fact that an old, festering wound has been reopened, namely, the ancient quarrel over "forbidden knowledge" and the profanation of sacred nature. Human curiosity in the scientific enterprise has gone too far; we have invaded and violated the sacred regions of birth and death. And now we are fated to reap the tragic consequences of such overweening curiosity. Our high, hard technology represents a nemesis, a fateful retribution for our having invaded the realm of forbidden knowledge.

But in the view of many others, this state of affairs cannot be adequately accounted for as a resuscitation of the quarrel over forbidden knowledge. Wolf Häfele suggests that the wide-ranging problems and passions raised by nuclear technology should be analyzed in light of one central realization: it is functioning in our society as a prototype, a forerunner, a pathfinder, compelling us to become far more constructive and imaginative in the task of dealing with an entirely new horizon of human possibility, hence a new order of creative risk. He writes:

> If properly interpreted and understood, the public concern about nuclear power is not unfounded. But that concern is not a simple function of a peculiarity of nuclear power. It is, rather, the general condition of civilization towards which we are moving; it is a condition where the magnitude of human enterprises becomes comparable with the magnitude of the widest determinants of our normal existence. Nuclear power turns out to be a forerunner, a pathfinder, of that.[7]

If we were sufficiently aware of the implications of Häfele's interpretation, we might get beyond at least two misconceptions.

The first is that we are still living in a society characterized by what C. P. Snow called, twenty years ago, "two cultures."[8] On the one hand scientists and engineers dominated by a "functionalist rationality" are trained within a universe of discourse dictated by the physical nature and limits of things. The risks of public safety which they perceive and try to minimize are derived from considering actually achievable technical options.

On the other hand, the humanist or philosopher or social reformer functions in a universe of discourse dominated by a philosophical vision of how things ought to be—quite apart from, even in spite of, the physical nature of technological possibilities and constraints. This vision leads to a negative perception of seemingly uncontrollable risks from powerful, complex energy systems which appear to take on a life of their own as they give aid and comfort to what many regard as man's myopic rape of the earth. The widening gap between these two universes appears to account for our rancorous dispute over an acceptable energy future.

But it seems closer to the facts to realize that we are now living in a Three Cultures society. While intellectuals and academicians carry on abstract, theoretical arguments about technological man and society, activists who have mastered the art of politics and media manipulation are engaged in a decisive struggle over the power to set a political agenda for an emerging era of profound social change. It would be a grave mistake for us to go on living an anachronism.

Secondly, if we were to realize the full implications of Häfele's observation, we might cease to regard nuclear science and technology as embodying some reprehensible Faustian bargain, or a transhuman inexorable Fate, and reëxamine it as the embodiment of highly evolved human values. To be sure, the amplification of human powers through science and its application to human needs is not an unmixed blessing, but neither is it an alien, vacuous, fateful fact divorced from the motivating values which initiate any scientific enterprise in the first place. Nuclear science and technology happen to be the embodiment of profound human values. Otherwise several who now live in the "atomic age" would not be so vigorously opposed to it in their strategy to arraign and all but convict nuclear technology as a scapegoat for less visible value-conflicts over a morally acceptable energy future.

If we expect to get beyond the rhetorical dissimulation in appeals to Faust, Fate, and Incantations, we must all endeavor to engage in the political process and to realize the full implications of Häfele's statements. The "general condition of civilization towards which we are moving" is fraught with perils and promises. One of its promises is that our human powers and sensitivity have reached a level of largesse that can implement conditions for providing a basic level of life-sustaining material well-being to everyone on this planet thereby mitigating in some measure the domestic unrest and instability that are the stuff of violence and wars. One of its perils is that we will fail to cultivate a geopolitical understanding of the consequences of our domestic political decisions. We will fail to become global citizens and instead retreat into our "local bastions of privilege."[8] We must find ways to make our public dispute over nuclear technology play a pathfinder role in our search for global political institutions that can effectively govern the enormous benefits of nuclear technology applied in medicine and electricity generation, and at the same time govern its potentially harmful abuses.

If anti-nuclear coalitions succeed in eliminating the nuclear option in the United States, it does not follow that there are good and sound reasons for it. Indeed, currently popular reasoning may reflect—not the failure of a new technology to meet rising expectations for "safety" criteria—but the failure to develop an adequate analytical and ethical framework for testing contradictory claims, deriving principles for organizing scientific evidence, and assessing the likelihood of alleged social consequences.

The need for such an ethical framework has become increasingly clear as the socio-political consequences of anti-nuclear assertions and arguments have begun to surface. Several arguments have been developed to support the belief that nuclear technologies not only *can* but *must* be totally eliminated for the sake of human survival. In each case these arguments have evoked counter-arguments appealing to a significantly different scientific basis.

To illustrate the need for a more adequate ethical framework let us first examine contradictory claims and arguments that nuclear technology can be eliminated.

II. Nuclear Technology Can Be Eliminated

A diverse body of literature—cross pollinated by environmentalists, economists, philosophers, futurologists, and others—has been integrated by several authors into a program for social and political reform which would virtually eliminate "hard" technology, epitomized by nuclear reactors, as well as centralized energy systems.[10] This program of reform stands or falls on the claim that the United

States has two mutually exclusive sets of technical choices leading to mutually exclusive paths, only one of which will usher in a sustainable energy future.

A. Arguments

The argument for abandoning a "hard path" in favor of a so-called "soft path," and transition thereto, derives from three premises.

The first is that "soft" technologies are already mature, technically proven, economically irresistible, and commercially available on a scale commensurate with growing human demands for adequate, reliable energy supplies throughout the world. The counter-claim is that soft technologies are now enjoying the same research and development euphoria that civilian nuclear electricity enjoyed twenty-five years ago, and that wishful thinking does not produce certified results in the market place.

A second premise is that the financial cost of conventional "high" technologies and fuels far exceeds the cost of soft technologies, both in capital investment and long-term maintenance. The Energy Research Group has published figures which counter this claim: the actual capital cost of paying for the proposed soft path would be $3 trillion, or 5.5% of the GNP between 1975—2000, a prohibitive economic outlay. Ian Forbes suggests that "A 'moderate path' employing significant conservation and primarily conventional technologies would have a 1975—2000 capital cost of $1 to 1.5 trillion dollars."[11]

A third premise commending the soft path option is that transitional and soft technologies can sustain our highly complex, industrial society without diminishing our standard of living, experiencing any energy shortfalls or civil disruptions or adverse environmental impacts. Those who reject this claim point out that the United States is now a nation of 220 million, with a doubling time no longer that of 100 years, but a mere 35 years. Demographers expect an additional 20 million new households by the year 2000. If energy supplies do not at the very least keep pace with this growth of population, the United States faces a future of certain distributional deficits and serious social disruption (not to mention the future of developing nations of the Third and Fourth Worlds.) The prospect of replacing natural gas and oil with decentralized, neighborhood windmills, solar collectors, residential wood and coal stoves deployed throughout a vast fuel distribution network serving individual customers—has not even been assessed by an environmental impact study required by NEPA legislation.

A second argument advanced to sustain the belief that we *can* have a non-nuclear future sums up all manner of social complaints. It suggests that we can have unlimited, free, zero-risk solar energy if only we dismantle the corporations. The Sun Day Committee distributed an open letter in April, 1978, with these introductory paragraphs:

Who will own the future energy supply of our nation, and what kind of energy will it be?

Will the giant corporations that control all present energy sources extend their monopoly to the unlimited energy source of the future—solar power? Will solar development be taken over by the energy giants and postponed until they've squeezed the last drop of profit from their investments in fossil fuels and nuclear power?

Will we be forever reliant upon a massive centralized energy system and eventually forced to pay for the use of our own sun?

Or will we have a future of unlimited, inexpensive, nonpolluting and decentralized solar energy?[12]

This document has prompted Professor George Pickering to observe that "there is more truth and less inuendo in a Mobil Oil ad than there is in that." He rejects the charge that there is some corporate conspiracy which keeps solar energy expensive, and that decentralizing it or legislating divestitute of potential corporate investors, will render it unlimited, nonpolluting and "free."[13]

The popular onsite philosophy so in vogue today, presumably ridding us of centralized utilities, has also been called into serious question. Marjorie and Aden Meinel, prominent in solar research for over twenty-five years, have insisted that solar systems suffer from five vexing problems: high capital costs, undependability of sunshine; the high cost of expensive, idle energy backup systems, unreliability of system-component lifetimes, and over-optimistic estimates of performance.[14]

Dr. Seymour Baron has made a net energy analysis of solar energy applications. He demonstrates that technologies presently under development are far from achieving anticipated degrees of energy conservation, "free sunshine," and environmentally benign conversion. "Both solar heating and thermal electric systems use large quantities of steel, concrete, aluminum, copper, plastic, and glass. Photovoltaic conversion to electric power is very energy-intensive when reducing silica to high purity silicon."[15] As solar technology now stands, the cumulative consumption of so-called non-renewable resources—including the oil, coal, uranium, and other fuels that must be consumed to build and operate solar energy systems—far exceeds the recovered energy. The energy required for producing just a few solar materials reach these totals:[16] 75 million BTU per ton of aluminum, 37 million BTU per ton of steel, 18 million BTU per ton of glass, 12 million BTU per ton of concrete.

B. Proposed Bioethical Framework

A fundamental bioethical principle is required to test contradictory claims that

(a) soft path technologies in general are technically proven, commercially available, less costly, more than adequate substitutes for current projected nuclear energy facilities; and

(b) solar technology in particular is a commercially reliable, less costly, non-polluting, risk-free, renewable substitute for current and projected nuclear technologies.

Ostensibly, the ethical problem about such contradictory claims is one of discerning the credibility (i.e., honesty, professional credentials, motives) of "competing experts." But to reduce the profound implications of the problem to a search for ethical criteria for credibility is at best superficial.

The substantive bioethical problem must be posed in very different terms. Which set of competing claims—if its technical promise and social consequences prove it to be false—would most seriously violate this fundamental bioethical principle?

Social justice requires an energy policy for the development of those technologies which have the potential—based on historical experience and verifiable scientific data—of preventing unjustifiable basic harm to the maximum number of human beings.

The classical admonition, "Do no harm," was formulated long ago as a directive for medical professionals in their conduct toward individual patients. The bioethical principle proposed here reflects an inescapable social reality, namely, that a public policy cannot possibly "do no harm." Any social policy will entail risks of some harm to some individual in some future circumstance.

A proposed social policy should derive its bioethical justification—not from the fact that tangible benefits for the greatest number of people can be demonstrated to result from it (the utilitarian justification)—but rather from the fact that a policy avoids unjustifiable *basic harm* for the maximum number of human beings. Basic harm results from a deprivation of basic goods essential for the material well-being of living human beings as a necessary condition for protecting both the biosphere and well-being of future generations (e.g., shelter, nourishing food, health, energy sufficiency, jobs, freedom of life choices).

If energy policy makers in the United States become persuaded that claims made about solar and soft technologies are valid, and that present projected nuclear technologies therefore can be eliminated, the test for moral responsibility will be the extent and degree to which soft technologies avoid widespread basic harm. If, however, expectations about solar and soft technologies cannot be met in reality, and if institutional obstacles cannot be overcome as rapidly as claimed, then the test of moral responsibility will be the extent and degree to which the use of nuclear technologies avoids widespread basic harm.

This is the analytic framework in which historical and scientific evidence should be organized. Debates about conflicting standards for a preferred "quality of life" are secondary to the more fundamental problem of avoiding basic harm by maintaining access to basic goods for the maximum number of living human beings.

These considerations introduce arguments advanced by those who claim that nuclear technology must be eliminated.

III. Nuclear Technology Must Be Eliminated

The bioethical principle proposed here expresses a recognition that there are kinds of human actions which inflict basic harm of such a nature that, no matter what benefits may also result, (e.g., a life-style closer to nature, cultural advantages of urban living) such actions cannot be justified. Doubtless the institution and practice of human slavery resulted in a tangible and advantageous benefits to slave owners, slave traders, and in some cases, to those enslaved. Nevertheless, the practice has merited moral and ethical condemnation because it inflicts basic harm by violating fundamental goods and values—namely, the essential human conditions of freedom and basic material well-being.

There are impassioned arguments that nuclear technologies ought to be eliminated because they are morally wrong, hence ethically unjustifiable.

A. Arguments

The moral argument for eliminating nuclear technologies —whether for electricity generation or medical diagnosis and theraphy—derives from the following line of reasoning.[17]

The public must be protected from all known carcinogens and mutagens. Radiation is generally regarded as a known carcinogenic and mutagenic agent. Because somatic and genetic effects from any level of radiation exposure are harmful to present and future

73

progeny, nuclear technologies impose biological and psychological harm on unconsenting citizens. Hence they are by their very nature immoral and unethical. Public policy, concludes the argument, should reflect the popular view that whatever can cause cancer or mutations probably will cause both, and therefore it must eliminate any commercial use of radiation sources as morally unjustifiable.

When the moral argument for eliminating nuclear technology is generalized into a principle, its policy implications become clearer: Anything that can cause harm to human health and the biosphere probably will cause harm; therefore, any use of a harmful agent is morally unjustifiable and must be prohibited by public policy.

Those who reject this line of reasoning question its fundamental tenet. Both the moral argument and its underlying principle contain a specious assumption, namely, that if you do not have exposure to radiation, you will not get cancer; but if you do have radiation exposure, you will certainly suffer from cancer and genetic mutations. Such reasoning is absurd.

Based on such principle and its policy implications, we would be bound to conclude that because all technologies—whether for producing energy or automobiles or airplanes—can and will cause harm, they are immoral and must be eliminated.

Critics consider it self-evident that, not only the moral argument, but also its underlying principle lead to an absurd position—one which negates rational discussion, destroys moral seriousness, and paralyzes the policy-making process. Clearly, reasoning within such an ethical framework is invalid.

B. Proposed Bioethical Framework

Again the primary ethical problem is not whether these arguments are being made by persons with scientific credentials or who have gained public credibility for whatever reason. The substantive question is whether the ethical framework in which interpretations of scientific data become organized and translated into moral arguments is intellectually sound and adequate.

From a bioethical perspective present and future generations will suffer a grave social injustice if serious moral concern for the life-sustaining qualities of the biosphere is allowed to become narrowed down to, and perhaps even trivialized by, an unwarranted preoccupation with perceived risks of adverse biological health effects from but one energy technology.

To protect the common good from this possibility, another fundamental bioethical principle is required:

74

Social justice and equity require a public policy that will protect citizens from potential sources of basic harm by their equitable management, that is to say, management which is proportional to actual, identifiable basic harms that human effort, time and money can reduce.

To implement this principle, a policy maker would be required to:

a) evaluate the entire spectrum of both natural and man-induced biohazards from energy alternatives;

b) make cost-comparisons of the available methods for per capita reduction of these various hazards, giving priority consideration to those that are certain in contrast to probable and merely possible;

c) only then make policies and set standards, based on evaluations, that will get the most public health protection for the most people out of a finite amount of money.

Implied in the bioethical principle of an equitable management of risks and potential harm is the judgment that no human activity whatsoever can qualify for the abstract category of being "risk-free" or "absolutely safe." Indeed a profound misconception of "safety" in popular literature dominates the controversy over public health protection. Safety is not an intrinsic, absolute, cost-free property that a given system or product or activity can and should possess. Neither is safety measured, much less predetermined, by the presence or absence of risks. Safety is an evolving value judgment derived from and related to changing personal or social priorities on a scale of real affordable possibilities.

In any viable social order, because of an unpredictable spectrum of biological and mental differences, some kinds of harm cannot be avoided. There are other kinds of harm which social policy should not permit. Ultimately, the bioethical problem of discriminating the difference comes down to a decision by policy-makers as to which harms are unavoidable and negligible by comparison with those harms which are much greater and avoidable, hence unjustifiable.

In contrast to a moral principle of absolute prohibition, the bioethical principle of "equitable management" of potential harms implies that judgments of safety ought to be value judgments about the justifiability or unjustifiability of harm. Within this bioethical framework, a logically consistent process of reasoning ought to be guided—not by a goal of zero-risk (a costly illusion)—but by comprehensive, comparative, risk/risk analyses and cost/risk/benefit ratios. When these comprehensive comparisons make it clear that a point of diminishing returns on allocations of human effort, time and money has been reached by comparison with allocation for other potential hazards to a population, then the particular product or process under scrutiny by policy makers is "safe enough." If unintended and unwanted harm should occur despite morally responsible policy making, then such harm can be judged ethically justifiable because unavoidable and negligible by comparison with other greater harms. With this process of reasoning, moral responsibility to present and future generations can have optimal expression.

Within this bioethical framework, it is neither morally nor ethically justifiable to single out for exclusive attention or even elimination one technology—nuclear fission—as the embodiment of moral disvalue, based on a single-minded consideration, namely that its use may entail risks of potentially adverse biological effects. The biological well-being of individuals is certainly a basic good to be protected, but not to the jeopardy of other basic goods essential to the well-being of an entire population. Protection of social well-being requires access to a wide spectrum of basic goods. Biological integrity is but one basic good, essential but not sufficient. Indeed what good is biological integrity if the means to nurture and develop its quality are missing or denied by social policy?

In a bioethical framework requiring an equitable management of all potential hazards as a moral imperative for public health and safety, another unresolved controversy needs to be considered—namely, whether or not there is a threshold for radiation exposure below which "no harmful effect" occurs.

The normative concept of a threshold has been accepted for most toxic elements, and it implies that below a threshold dose, any exposure is "absolutely safe." In the course of evolving radiation protection standards, however, the state-of-the-art and prudence have induced policymakers to adopt a conservative assumption: better to assume some harmful effect from any radiation, however small, than to assume a threshold dose and then discover data to the contrary. Consequently a no-threshold, linear hypothesis has been used by regulators in setting and enforcing standards. But the ensuing quest for zero-exposure to radiation as low as reasonably achievable (ALARA) rests only on an unproven hypothesis, a conservative assumption, and not on a scientifically established fact.

Since we do not know the effects of low-level radiation exposures, and since human beings have been accustomed to experiencing wide variations in exposure to natural background radiation, we must ask what scientific evidence can justify ever more costly, marginal reductions of exposure if "conservative" actually means doing the least harm? J. J. Cohen observes:

"Supposedly any degree of reduction in radiation exposure will do some good. However, some evidence indicates that there might in fact be a *net beneficial effect* of radiation at low levels. Since we do not in fact have a complete understanding of low-level exposure phenomenologically, perhaps we should recognize the possibility of beneficial as well as harmful effects. If the net effects are in fact beneficial, then by insisting upon the application of ALARA—rather than being conservative we may actually be causing harm.[18]

Since scientific evidence demonstrates net beneficial effects of low-level exposure to other toxic elements (e.g., copper, selenium, fluoride), professional ethics require scientists to examine the possibility that there may also be net beneficial effects from exposure to low-levels of radiation. Until the results of this research endeavor into positive effects as well as negative effects has been completed, there is no bioethical justification for more costly and marginal reductions of radiation exposures previously judged "safe."

The bioethical principle of an equitable management of potential biohazards further suggests the urgent need for an underlying "philosophy of congruence" with a pattern of benefits and harms already established by naturally occurring sources of radiation and other toxic elements. Human beings have been evolving and increasing their life expectancy within a background of naturally occurring toxic and radiation sources throughout recorded history.

Large segments of the United States population receive natural external radiation doses varying from 40 to 150 millirem per year simply because of geographic location. Natural exposure to thorium in monazite sands along the southeastern coast of India varies from 130 millirem to 2,800 millirem. On the coast of Brazil radiation exposures vary from 90 millirem to 2,800 millirem; the average is 550 millirem per year. There is no scientifically established evidence that the people exposed to these wide variations suffer basic harm.

Human tolerance for, indeed dependence upon, such wide variations in natural radiation exposure for several millenia demonstrate that any increments from man-made applications of natural sources can be kept well within the range of those variations without inflicting either unjustifiable harm or deprivation of basic goods to members of society.

The bioethical principles set forth above, and an underlying philosophy of congruence, suggest a third bioethical principle to guide formulation and provide moral justification for social policies protecting health and safety.

Any involuntary risks imposed by social policies for protection against radiation and other toxic elements must be congruent with, must not be in excess of, and may be reasonably less than, those involuntary risks imposed by the wide variations in naturally occuring toxic elements and harmful effects from our natural environment.

An objective application of this principle can be tested by examining proposed performance criteria for the management and disposal of radioactive wastes. These criteria propose that ultimate waste disposal shall be conducted in such a manner that there is no net increase in risk of harm to people or the biosphere by comparison with the typical ore body of natural uranium which yields the energy from which wastes are derived. In other words, disposal methods for wastes would return them to the same (or reasonably better) level of risk already posed by natural uranium ore in the earth's crust. According to this concept, technology would be required to make the waste form have at least the same stability as the original ore body. The medium containing the waste would be required to retain at least the same integrity as the medium containing the ore. The geological media surrounding and isolating the waste would be required to retain the same integrity of isolation from the biosphere as the isolation media of original ore bodies.

Any risks of adverse health effects both to present and future generations from radioactive wastes should be measured—not by the length of their toxic lifetimes—but only in relation to environmental pathaways which determine the degree of likelihood of harmful exposure of and assimilation by the human body. All of the pathways analyses to date have measured those risks and found them to be vanishingly small.[19] Moreover, scientists have made a calculation considering only eight toxic elements naturally occurring in the earth's crust and continuously leached into food and water supplies which we daily use. Those elements are mercury, lead, cadmium, chromium, selenium, barium, arsenic, and uranium. If the entire electrical supply of the United States for 100 years came from nuclear fissioning of uranium, and the wastes were buried on the earth, the resulting increase in the toxicity of the earth's crust would be one ten-millionth of one per cent. Those who wish to invalidate that comparison say that the other toxic elements are distributed more uniformly, but the waste would be concentrated in a few repositories. The fact is that nature has also concentrated toxic minerals in ore bodies. Cohen states:

> It can be shown that in a few hundred years the repository contents become relatively less toxic than typical mercury deposits and in about 1000 years it becomes less than the uranium ore body from which the nuclear fuel was originally derived.

The ore body is at least as available to dissolution and transport as is the waste repository.[20]

From a bioethical perspective, if technologies exist to meet performance criteria outlined above, as well as a normative philosophy of congruence with a natural pattern of benefits and harms, then a moral and ethical justification exists to implement these criteria.

IV. The Inevitable Fact—A Nuclear Future

Someone has observed that electricity itself would still be on trial if the electric chair had been invented before the light bulb. It appears that the public's highly selective fear of nuclear electricity, and general nuclear paranoia, has its roots in "the Bomb", together with associated feelings of guilt and apprehension about a future thermonuclear war. Accordingly, the problem of proliferation of nuclear weaponry seems so intractable that only their elimination assures protection.

Several political activists insist that "the Bomb" and nuclear electricity are as inseparable as Siamese twins. Apparently current presidential policy has accepted this unexamined belief. President Carter stated publicly that by setting a "moral example" to the rest of the world in our moratorium on reprocessing spent fuel, further development of breeder technology, and embargoes on export of any other type of nuclear technology, the United States can unilaterally and singlehandedly "put the nuclear genie back in the bottle."

Preventing the spread of nuclear weapons is a laudable intention, but it suffers from two deficiencies. First it assumes that the mere existence of weapons is a cause of war. Secondly, it assumes that electricity-generating reactors are the most direct, cheapest and attractive way to get such weapons. In fact, present policy sets up global conditions that will encourage the rapid spread of nuclear technology in the wake of fierce international competition over oil in diminishing supply and at an escalating price—and conflicts over fuels to sustain that technology—without the safeguards that would be exchanged for a timely and reliable access to nuclear technology and fuels.

The naive belief that the world can and must have a non-nuclear future has become an academic exercise, and the gap between charismatic rhetoric and political realities is growing ever wider. The world is going to have a nuclear future, and this means that policy decisions are going to be made for good, or for ill. It grows clearer every day that this nation's leaders have already sacrificed too much of our domestic and global nuclear policy on the altar of nonproliferation.

A far more enlightened policy for the United States would be to assume leadership in the development of internationally controlled and safeguarded civilian nuclear fuel-cycle facilities—including spent fuel temporary storage, reprocessing, and all handling or transport operations involving weapons usable material.[21] Such facilities could be required to remain subject to international on-site inspection and security control. The economic incentive for regional centers and international cooperation will continue to exist for some time to come.

However, if the United States fails to assume the responsibility for this leadership—if we lose the window we now have for entering the arena of political innovation—then our nuclear future on this globe will be significantly different, and fraught with dangers that are here and now avoidable. Failing that leadership, isolated nationalistic needs of each country will dictate the development and deployment of civilian nuclear technology. To assure security, civilian operations would likely come under military surveillance. Only token levels of international inspection and control would be accepted—if any at all. Nuclear technology would then probably never become demilitarized, but progressively the opposite. It uses would become a military property, withdrawn from civilian control and oversight. Socially beneficial uses would be dictated by military goals.

But, of course, there is an enormous ideological attraction to the assumption that somehow a "phasing out" of all forms of nuclear technology and weaponry can and will occur worldwide. To the idealist, this must be attempted—no matter how improbable and hazardous. We have already seen this ideal at work by militant political activists—storming sites of plant construction—and career intervenors initiating interminable legal delays and embargoes on reprocessing and the breeder. These political actions have all been designed with the hope that they will so disrupt the supply-and-demand balance that hundreds of reactors, tons of plutonium (already existing in weapons and spent fuel) and millions of gallons of military radwastes will somehow disappear.

Unfortunately, it is folly to believe that massive collective amnesia, or Great Renunciation, or Divine Fiat will make nuclear science and technology go away. Such naive idealism only escalates the likelihood of international instability, conflict over diminishing fossil fuels, and the possibility of the very war that idealists strain to avoid.

Those who wishfully believe that solar power is a technological fix that will deliver us into a non-nuclear future and obviate the need to confront the political problem of devising institutions to govern nuclear technology—or that solar power will not itself be harnessed for destructive weaponry when it becomes technically feasible to do so—live in a world of make-believe. The political problem will perdure—namely, to devise institutions and safeguards on a scale of resolve and human ingenuity yet unknown. If we do not learn to govern nuclear science and technology, we cannot hope to govern other potential sources of weaponry and this is a primary condition under which a world without war is a possible, politically meaningful goal. We must also recognize that the most paralyzing, debilitating, and manipulable human emotion is fear—fear begotten of ignorance, intellectual laziness, and moral superficiality.

Notes

1 Amory B. Lovins, SOFT ENERGY PATHS: TOWARD A DURABLE PEACE (Cambridge, Mass.: Ballinger Publishing Co., 1977)
2 Amory B. Lovins, L. Hunter Lovins, Leonard Ross, "Nuclear Power and Nuclear Bombs," *Foreign Affairs* (Summer 1980) pp. 1137-1177.
3 Irvin C. Bupp & Jean-Claude Derian, LIGHT WATER: HOW THE NUCLEAR DREAM DISSOLVED. (New York: Basic Books, 1978)
4 Text reported in NUCLEAR LIGHT AND POWER (24 March 1975) Cf. also: "Our Lives May Be At Stake," in REDBOOK (November 1974)
5 Peter L. Berger, "Ethics and the Present Class Struggle," WORDVIEW (20 April 1978) pp. 6-11. Also, Llewellyn King, "Nuclear Power in Crisis: The New Class Assault," ENERGY DAILY, 6/ # 135 (14 July 1978). G. W. Pickering, "Science and Society in the 70's: The Making of a New Agenda," *Proceedings of the Fifth Life Sciences Symposium*, Los Alamos, 1978.
6 Alvin Weinberg, "Social Institutions and Nuclear Energy," SCIENCE, 177 (7 July 1972)
7 Wolf Häfele, "Hypotheticality and the New Challenges: The Pathfinder Roge of Nuclear Energy," MINERVA, 12 (1974) 303-322.
8 C. P. Snow, THE TWO CULTURES AND THE SCIENTIFIC REVOLUTION (New York: Cambrcdge Univ. Prds, 1959)
9 G. W. Pickering, "Ohe Road Not Taken—And Wisely So: A Path Too Soft to Travel," ELECTRIC PERSPECTIVES 77/3, p. 7
10 cf. Lovings *supra* n. 1. Also Wendell Berry, *The Unsettling of America*, San Francisco, Sierra Club Books, 1977; John Todd, *The Future Is Our Permanent Address*, (New York: Harper and Row, 1978).
Barry Commoner, *The Politics of Energy*, (New York: Alfred A. Knopf, 1979).
11 Ian A. Forbes, "The Economics of Amory Lovins' Soft Path," Energy Research Group, Waltham, Mass., 1978
12 Letter signed by Denis Hayes, Chairman, Senior Researcher, Worldwatch Institute, Washington, D.C., 1978
13 cf. Pickering, *supra* No. 5
14 A. Meinel and M. Meinel, "Hard Realities of the Soft Path to a Solar Future," in PROCEEDINGS of the Second INTERNATIONAL SCIENTIFIC FORUM ON AN ACCEPTABLE WORLD ENERGY FUTURE, Nov. 27—Dec. 1, 1978. (Ballinger, in press).
15 S. Baron, "Solar Energy—Will it Conserve Our Non-Renewable Resources?" THE PUBLIC UTILITIES FORTNIGHTLY, 28 Sept. 1978
16 P. Beckmann, "Why 'Soft' Technology Will Not Be America's Energy Salvation," Different Drummer Booklet No. 6 (Boulder, Colo.: Golem Press, 1979)

17   Dr. John Gofman's version of this moral argument is popularly cited and accepted. It appears in unpublished manuscript form in an essay entitled "The Question of Law Versus Justice," date December 2, 1978.

18   J. J. Cohen, Personal Communication

19   There are four peer-reviewed pathways analyses:

    1) B. L. Cohen, "The Disposal of Radioactive Wastes from Fission Reactors," SCIENTIFIC AMERICAN, 236 (June, 1977)

    2) G. de Marsily et al., "Nuclear Waste Disposal: Can the Geologist Guarantee Isolation," SCIENCE, 197 (August 1977)

    3) "Report to the American Physical Society by the Study Group on Fuel Cycles and Waste Management," REVIEW OF MODERN PHYSICS, 50/Part II (Jan. 1978)

    4) "Handling of Spent Nuclear Fuel and Final Storage of Vitrified High Level Reprocessing Waste, KBS REPORT—available from KarnBransle-Sakerhet, Stockholm, Sweden (December 1977)

20   J. J. Cohen, "Public Testimony before the Interagency Review Group on Nuclear Waste Management," San Francisco, CA, Public Hearings, 21 July 1978

21   The options outlined here are summarized from E. L. Zebroski, "Proliferation Risk Management for an Acceptable Nuclear Energy Future," in NUCLEAR ENERGY AND ALTERNATIVES. Proceedings of the International Scientific Forum on Acceptable Nuclear Energy Future. (Cambridge, Mass.: Ballinger, 1978)

V

ABOUT THE AUTHORS. . .

Andrew Hacker is a Professor of Political Science at Queens College, City University of New York, and is the author of numerous books and articles on American institutions and politics. From 1955 to 1971, Mr. Hacker was an Assistant Professor, Associate Professor and Professor in the Department of Government at Cornell University.

He holds degrees from Amherst College (A.B., 1951), Oxford University (M.A., 1953), and Princeton University (Ph.D., 1955), and has taught at Queens College since 1971.

His books include: *Political Theory: Philosophy, Ideology, Science* (Macmillian, 1961); *The Study of Politics* (McGraw-Hill, 1963, 1973); *Congressional Districting* (Brookings Institution, 1964); *The End of an American Era* (Athenaeum, 1970); *The New Yorkers* (Mason-Charter, 1975); and *Free Enterprise in America* (Harcourt, Brace, 1977).

Robert L. DuPont is a practicing psychiatrist and president of the non-profit Institute for Behavior and Health, Inc. in Rockville, Md. He is a diplomat of the American Psychiatry and Neurology, a fellow of the American Board of Psychiatric Association, and a member of many professional organizations, including the Academy of Behavior Medicine Research, the Behavioral Medicine Special Interest Group, the American Public Health Association, and the District of Columbia Medical Society. He is also chairman of the Drug Dependence Section of the World Psychiatric Association, a position he has held since 1974. He is a Visiting Associate Clinical Professor of Psychiatry at Harvard Medical School.

Dr. DuPont is a graduate of Emory University, and received his M. D. from Harvard Medical School. He was medical intern at Cleveland Metropolitan General Hospital, Western Reserve

Medical School; psychiatric resident and teaching fellow, Massachusetts Mental Hospital Center, Harvard Medical School; and Clinical Associate, National Institute of Health.

In May, 1978, Dr. DuPont chaired a Special Session at the American Psychiatric Association's (APA) Annual Meeting in Atlanta on the "Treatment of Phobia." He chaired a similar Special Session at the 1970 APA meeting in Chicago. In 1980 he led the second annual National Phobia Conference held in Washington, D. C.

He has authored mord than 100 professional articles and a book on topics in the field of health promotion, drug abuse prevention, and criminal justice. In addition, he appears on ABC-TV's "Good Morning America" program and has contributed to local and national TV, radio, magazines, and newspapers on a variety of health topics.

William Barrett is a Professor of Philosophy at New York University who describes his main areas of concentration as the History of Philosophy, Existentialism and Phenomenology, and Aesthetics.

Mr. Barrett, a native New Yorker, attended New York schools and earned degrees from City College (B. A., 1933) and Columbia University (M. A., 1934, Ph.D., 1939). He holds numerous postdoctoral awards and distinctions and has taught at New York University since 1950.

He is the author of, among other works: *Aristotle's Theory of Movement* (Columbia University Press, 1939); *What is Existentialism?* (Partisan Review Monograph, 1947); *Zen Buddhism* (with D. T. Suzuki, Doubleday, 1956); *Irrational Man* (Doubleday, 1958); *Philosophy in the Twentieth Century* (with Henry Aiken, Random House, 1962, reprinted by Harper & Row, 1970); *What is Existentialism?* (Grove Press, 1964); *Ego and Instinct* (with D. Yankelovich, Random House, 1970); *Time of Need: Forms of Imagination in the 20th Century* (Harper & Row, 1972); and *The Illusion of Technique* (Doubleday, 1978).

Margaret Maxey is Assistant Director of the South Carolina Energy Research Institute in Columbia, S. C. Prior to her appointment she was Associate Professor of Bioethics at the University of Detroit. She also has served as Assistant Professor of Philosophy and Religion at Barat College, Lake Forest, Illinois.

Ms. Maxey received her B. A. degree in philosophy at the Creighton University, Omaha, and M. A. in philosophy (1963) at St. Louis University, and a second M. A. degree, in systematic theology (1967) at the University of San Francisco. Her Ph.D. degree in

84

Christian Ethics was completed at Union Theological Seminary, New York, in 1971.

Her publications include:

"The National Council of Churches and Nuclear Power: A Response," *Christianity and Crisis,* 10 May 1976

"Nuclear Energy Debates: Liberation or Development?" *The Christian Century,* 21/28 July 1976

"Public Ethics and Radioactive Wastes: Criteria for Environmental Criteria," *Proceedings of a Workshop on Issues Pertinent to the Development of Environmental Protection Criteria for Radioactive Wastes,* U. S. Environmental Protection Agency, Washington, D. C., 1977

"Along the Soft Path to Soft Technologies: Cofts and Casualties," *Electric Perspectives* (No. 77/3), June, 1977

"Nuclear Energy Politics: Moralism vs. Ethics," *Ethics and Public Policy Reprint No. 1;* Ethics and Public Policy Center, Georgetown University, 1977

"A Bioethical Perspective on Acceptable-Risk Criteria for Nuclear-Waste Management," University of California, Livermore: UCRL-52320, July 15, 1977

"Hazards of Solid Waste Management: Bioethical Problems, Principles, and Priorities," *Proceedings of Fifth Life Sciences Symposium,* Los Alamos Scientific Laboratory, Los Alamos, New Mexico, October, 1977

"Radwastes and Public Ethics: Issues and Imperatives," *Health Physics,* 34/2, February, 1978

"Radiation Protection Philosophy: Bioethical Problems and Priorities," *American Industrial Hygiene Association Journal,* September, 1978

"Energy, Society, the Environment: Conflict or Compromise?" *Ethics and Energy,* Washington, D. C., Edison Electric Institute, 1979

"Radiation Health Protection and Risk Assessment: Bioethical Considerations," *Perceptions of Risk:* Proceedings of Fifteenth Annual Meeting of the National Council on Radiation Protection and Measurements, Washington, D. C.; National Academy of Sciences, 1979. pp. 18-32.

MORE INFORMATION?

First, contact your local electric company. You'll find they have a wealth of information in their own publications and in other EEI documents.

Other volumes of *Decisionmakers Bookshelf*

To obtain Decisionmakers Bookshelf volumes, write Order Department, Edison Electric Institute, 1111 19th Street N.W., Washington, DC 20036. Price of individual volumes, $2.50.